Large-Scale Construction Project Management

Large-Scale Construction Project Management

Understanding Legal and Contract Requirements

Yan Tan

CRC Press is an imprint of the
Taylor & Francis Group, an **informa** business

First edition published 2020
by CRC Press
6000 Broken Sound Parkway NW, Suite 300, Boca Raton, FL 33487-2742

and by CRC Press
2 Park Square, Milton Park, Abingdon, Oxon, OX14 4RN

CRC Press is an imprint of Taylor & Francis Group, LLC

ISBN: 978-1-138-38933-5 (hbk)
ISBN: 978-0-429-42399-4 (ebk)

Typeset in Times
by Deanta Global Publishing Services, Chennai, India

Contents

Preface

For decades, large-scale construction projects have had a poor reputation for delivery within the budgeted cost and planned completion date. Although research continues into various aspects of improving management of construction projects, in reality there have been barely any improvements so far.

For the past decade, I gained experience and skills working with various industry-leading firms and professional organizations to explore the effective tools and methods to improve large-scale construction project management. I have been involved in many major construction projects worldwide. I saw which key success factors of the project went well and learned the lessons when things went wrong. For all large-scale construction projects, in order to achieve project success, it is very important for the parties to understand the contractual and legal requirements and establish an integrated project control system at the front-end of the project.

In recent years, the leading standard forms of contract all published their new editions, e.g., JCT 2016, NEC4 and FIDIC rainbow suite second edition in 2017. In addition, PRINCE2 published new guidance in 2017, and PMI has updated a number of its standards and guides since 2017, including the sixth edition of the famous PMBOK in 2017, the new "Construction Extension to the PMBOK® Guide" in 2016, the fourth edition of the "Standard for Portfolio Management" in 2017, the fourth edition of the "Standard for Programme Management" in 2017, the third edition of the "Practice Standard for Scheduling" in 2019, the third edition of the "Practice Standard for Work Breakdown Structures" in 2019, and the "Standard for Risk Management in Portfolios, Programs, and Projects" in 2019.

I believe it is very useful to publish a book which integrates the relevant provisions of these leading contracts, together with the leading project management standards set out in PMBOK and PRINCE2, with the latest updates. With my broad expertise in project management, project planning, risk management, contract management, commercial management, construction law, and dispute resolutions, plus my extensive working experience with various world-leading organizations, I believe that I can establish the interlinks between the contract and legal requirements, with the leading project management tools and methodologies, and practical experience.

In addition, this book has also incorporated the latest developments in construction law associated with some burning issues, for example the concurrent delay and the "Smash and Grab" adjudication. In order to support continuous improvement of the project control after establishing the integrated project control system, Lean construction is also considered in this book.

The aim of this book is to facilitate improvements in the management and control of large-scale construction projects from the early stages of the project by understanding the legal and contractual requirements during the implementation of the large construction project. I also hope this book not only sets out a correct path for students undertaking project management courses, but also industry practitioners, e.g., project planners, commercial managers, contract administrators, project managers, risk managers, project control managers, and programme managers, to improve the control and management of their projects.

Following the success of the NEC and FIDIC contracts in delivery of large international construction projects, I hope this book can also help project management professionals outside of the UK to understand the concept of the latest version of these contracts. In particular, as regards forming the appropriate contract before commencement and establishing an effective integrated project control system for delivery.

Dr. Yan Tan
PhD MBA MSc BEng FCIArb PMP PMI-RMP NECReg PRINCE2-PR

Acknowledgments

I would like to express my gratitude to Oracle and its subsidiaries for giving me permission to use images of their products.

I am thankful to all reviewers who offered comments. Special thanks to Faiyaz Yunus from Jacobs Engineering, Nigel Hayward from Arcadis, and Phil Howe from Honeywell. Special thanks also to Joseph Clements and Lisa Wilford from Taylor & Francis Group, LLC for their assistance in the preparation of this book.

Thanks to my husband, Dr. Feng Fu, who inspired me to write this book in the first place. Thanks also to my family for their continuous support in finishing this book.

Author

Yan Tan, PhD, earned her PhD in Project Risk Management from the University of Leeds, UK, MBA from the University of Manchester, UK, and Master of Construction Law and Dispute Resolutions from King's College London, UK. Prior to these, she also earned a double Bachelor's degree in International Project Management and Electrical Engineering from Tianjin University, China.

Besides her academic achievements over a broad range of subjects, including Law, Finance, Risk Management, Project Management, Civil Engineering, Electrical Engineering, and Computer Science, Dr. Tan is a Project Management Institute (PMI) qualified Project Management Professional (PMP) and Risk Management Professional (PMI-RMP). She was Vice-Chair of the Construction Forum of the Project Management Institute (PMI) UK Chapter from 2015 to 2018. She is also a qualified PRINCE2 Practitioner and NEC accredited Project Manager (NECReg). In addition to her achievements in project management, Dr. Tan is a Fellow of the Chartered Institute of Arbitrators (FCIArb), a member of the Arbitration Club, and a Freeman of the City of London.

Dr. Tan has worked in a number of the world's leading organizations, including China Construction Bank, Jacobs Engineering, Arcadis, Honeywell, Vinci, Arup, and Capita, where she gained extensive experience in managing large-scale construction projects worldwide.

Abbreviations

5S	Sort, Set, Shine, Standardize, Sustain
AAA	American Arbitration Association
AACE	American Association of Cost Engineers
AC	Actual Cost
AC	Adjudication Contract
ACA	Association of Consultant Architects
ADB	Asian Development Bank
ADR	Alternative Dispute Resolution
AfP	Application for Payment
AHP	Analytic Hierarchy Process
AIA	American Institute of Architects
ALC	Alliance Contract
BAC	Budget at Completion
BCIS	Building Cost Information Service
BEI	Baseline Execution Index
BGB	Bürgerliches Gesetzbuch
BIM	Building Information Management
BIT	Bilateral Investment Treaties
BoQ	Bills of Quantity
CAP	Conflict Avoidance Panel
CE	Compensation Event
CEDR	Centre for Effective Dispute Resolution
CIArb	Chartered Institution of Arbitrators
CIC	Construction Industry Council
CIMAR	Construction Industry Model Arbitration Rules
CPI	Cost Performance Index
CPLI	Critical Path Length Index
CPM	Critical Path Management
CPRs	Civil Procedure Rules
CV	Cost Variance
DAAB	Dispute Avoidance and Adjudication Board
DAB	Dispute Avoidance Board
DB	Design and Build
DBO	Design Build Operate
DCMA	Defense Contract Management Agency
DMAIC	Define, Measure, Analyse, Improve, Control
DNP	Defects Notification Period
DRB	Dispute Review Board
DRBF	Dispute Resolution Board Foundation
DRSC	Dispute Resolution Service Contract
EAC	Estimate at Completion
EBRD	European Bank for Reconstruction and Development

ECC	Engineering Construction Contract
ECI	Early Contractor Involvement
EMV	Expected Monetary Value
ENE	Early Natural Evaluation
EOT	Extension of Time
ETC	Estimate to Complete
EWN	Early Warning Notice
FC	Framework Contract
FF	Finish to Finish
FIDIC	Fédération Internationale Des Ingénieurs-Conseils
FMEA	Failure Model Effect Analysis
FMEA	Failure Modes and Effects Analysis
FS	Finish to Start
HGCRA	Housing Grants, Construction and Regeneration Act
HOO	Home Office Overhead
IBRD	International Bank for Reconstruction and Development
ICC	International Chamber of Commerce
ICE	Institute of Civil Engineering
ICSID	International Centre for Settlement of Investment Disputes
JCT	Joint Contract Tribunal
JCT	Joint Contracts Tribunal
LCIA	London Court of International Arbitration
LHS	Latin Hypercube Sampling
LoE	Level of Effort
MDB	Multilateral Development Banks
NEC	New Engineering Contract
NED	Non-Excludable Delay
NOD	Notice of Dissatisfaction
NYC	New York Convention
ONS	Office of National Statistics
PERT	Programme Evaluation and Review Technique
PLN	Pay Less Notice
PMI	Project Management Institute
PMO	Project Management Office
PN	Payment Notice
PRA	Primavera Risk Analysis
PRINCE2	PRojects IN Controlled Environments
PSC	Professional Services Contract
PV	Planned Value
PWDD	Price for Work Done to Date
QCRA	Quantitative Cost Risk Analysis
QSRA	Quantitative Schedule Risk Analysis
RBS	Risk Breakdown Structure
RIBA	Royal Institute of British Architects
RICS	Royal Institution of Chartered Surveyors
SBC	Standard Building Contract

SC	Supply Contract
SCC	Schedule of Cost Components
SCC	Stockholm Chamber of Commerce
Scheme	Scheme for Construction Contracts (England and Wales) Regulations 1998
SCL	Society of Construction Law
SF	Start to Finish
SGA	Sales of Goods Act
SGSA	Supply of Goods and Services Act
SPI	Schedule Performance Index
SPV	Special Purpose Vehicle
SS	Start to Start
SV	Schedule Variance
SWOT	Strengths, Weaknesses, Opportunities, Threats
TCC	Technology and Construction Court
TCPI	To-Complete Performance Index
TSC	Term Service Contract
UNCITRAL	United Nation Commission on International Trade Law
VOB	Vergabe- und Vertragsordnung für Bauleistungen
WBS	Work Breakdown Structure

1 Introduction

1.1 OVERVIEW

Due to urbanization and population growth, the demand for infrastructure facilities keeps increasing. According to a report from leading economic organizations, there is a significant gap between demand and supply in global infrastructure investment. The World Economic Forum (2017) estimates the global investment demand for infrastructure projects is £3.7 trillion annually and currently there is an investment shortage of US$1 trillion needed to maintain the existing economic growth rate. In the United Union report, McKinsey & Company (2016) provide that emerging economies account for some 60 percent of such needed infrastructure investment. The Asian Development Bank (2016) estimates that the investment demand of Asian countries (excluding China) for infrastructure projects is up to 5% of GDP. In PwC's report "Capital project and infrastructure spending: outlook to 2025," Oxford Economics (2016) estimates that global infrastructure investment will double to US$9 trillion annually by 2025. According to the infrastructure report from the American Society of Civil Engineers (ASCE), following the assessment of the USA's infrastructure grade every four years, it emerged that the USA will need to invest US$4.59 trillion by 2025 to improve the nation's infrastructure (Business Insider, 2017). Surprisingly, Washington state has the worst road status with 89% of roads in poor condition (ASCE, 2017). It would require US$836 billion just to upgrade the roads in the USA (ASCE, 2017).

Due to the complexity and scale of these infrastructure projects, the challenges involved in the project life cycle are enormous. Budget overrun and schedule delay are common issues in managing and controlling large construction projects. If the issues are not resolved in time, it gives rise to disputes at a later stage.

Time and cost are the key elements of a construction claim. In practice, the common issues during project implementation may lead to disputes such as follows:

1) unbalanced risk allocation during contract draft
2) insufficient change management throughout the project implementation
3) lack of appropriate programme in place from project commencement
4) lack of effective integrated management and control system or methodology

All of the above issues create problems when managing and controlling large construction projects. It is important to understand the contractual and legal requirements for the project and establish integrated project management and a controlled approach at the front-end of the project. Such an approach can also facilitate early settlement of disputes and eliminate potential disputes at an early stage.

1.2 AIM AND OBJECTIVES

The aim of this book is to establish front-end dispute avoidance management and control framework for large construction projects. In order to achieve this aim, the following objectives are established:

- to analyze the provision requirements under standard forms contract
- to understand the relevant legal principle and applicable law
- to explore potential improvement of contract provisions in relation to time, cost, risk, and change management
- to understand the dispute resolution mechanism and explore how front-end project governance can be used for reducing disputes in advance
- to develop the integrated project control mechanism

1.3 CONTENT

This book starts by introducing the well-known international standard forms contract, including FIDIC, NEC, JCT, the history of their development, features of the contract, and recent changes in the latest editions.

The book then explains the requirements under different standard forms contract and legal principles for each knowledge area including time management, cost management, change management, and risk management in separate chapters respectively. Following thorough explanation and analysis, the contract requirements and practice tips and pitfalls, best practice as well as recommendations in each knowledge area conclude each chapter.

The following chapter introduces the well-used alternative dispute resolutions mechanisms in large international construction projects. The dispute resolution mechanism under each standard forms of contract will then be discussed. The legal requirements under different jurisdictions are explained. The advantages and disadvantages of relevant alternative dispute resolution approaches is considered. The recommendation of dispute resolution clauses during the contract drafting stage is then provided.

The book then introduces project control in an integrated manner, aiming to develop an effective integrated tool to effectively manage and control the cost, time, risk, change, and contract for large construction projects. Finally, the book concludes by recommending the establishment of effective dispute avoidance management and control framework in the front-end of large construction projects.

1.4 SCOPE AND LIMIT

Due to the recent release of new editions of all well-known standard forms contract and in the absence of further notes, this book will refer to the content under the most recent version of each standard forms of contract and refer to the common form of contract for each contract suite. Therefore, without additional notes, the following contracts will be used:

- FIDIC: Construction Contract Second Edition 2017 Red Book
- NEC: NEC4 Engineering and Construction Contract (ECC)
- JCT: Design and Build Contract 2016 (DB 2016)

Nevertheless, the author appreciates that the NEC3 and FIDIC 1999 editions are still the most popular standard forms of contract used in the market. Therefore, the author will specify the implementation differences in the implementation of NEC3 and FIDIC 1999 at key moments.

It worth noting that the identified terms are presented in italics and the defined terms start with capital letters under the NEC contract. Therefore, when specific terms or features are discussed, particularly for the NEC contract, the relevant terms will be presented in the relevant format as required in the NEC contract.

This book focuses on management and the control of time, cost, change, and risks in large construction projects. Other knowledge areas in project management, such as quality management, design management, and communication management are not within the scope of this book.

REFERENCES

2017 Infrastructure Report Card. ASCE. https://www.infrastructurereportcard.org/wp-content/uploads/2017/01/Roads-Final.pdf, Accessed on 15 July 2019.

Bridging Global Infrastructure Gaps. McKinsey & Company. http://www.un.org/pga/71/wp-content/uploads/sites/40/2017/06/Bridging-Global-Infrastructure-Gaps-Full-report-June-2016.pdf, Accessed on 15 July 2019.

Capital Project and Infrastructure Spending Outlook to 2025. PWC. https://www.pwc.com/gx/en/capital-projects-infrastructure/publications/cpi-outlook/assets/cpi-outlook-to-2025.pdf, Accessed on 15 July 2019.

Global Infrastructure Outlook. Global Infrastructure Hub. https://s3-ap-southeast-2.amazonaws.com/global-infrastructure-outlook/methodology/Global+Infrastructure+Outlook+-+July+2017.pdf, Accessed on 28 September 2019.

Infrastructure Outlook, Forecasting Infrastructure Investment Needs and Gaps. https://outlook.gihub.org/, Accessed on 28 September 2019.

Introduction: The Operations and Maintenance (O&M) Imperative: The Global Infrastructure Gap. World Economic Forum. http://reports.weforum.org/strategic-infrastructure-2014/introduction-the-operations-and-maintenance-om-imperative/the-global-infrastructure-gap/>, Accessed on 15 July 2019.

Meeting Asia's Infrastructure Needs. Asian Development Bank. https://www.adb.org/sites/default/files/publication/227496/special-report-infrastructure.pdf, Accessed on 15 July 2019.

Members and Prospective Members of the Bank. Asian Infrastructure Investment Bank. https://www.aiib.org/en/about-aiib/governance/members-of-bank/index.html, Accessed on 11 September 2019.

The US Will Need to Invest More Than $4.5 Trillion by 2025 to Fix Its Failing Infrastructure. Business Insider. http://uk.businessinsider.com/us-invest-over-4-trillion-by-2025-to-fix-infrastructure-2017-3?r=US&IR=T, Accessed on 15 July 2019.

2 Construction Contract

2.1 INTRODUCTION

A construction contract is the binding agreement between the Employer and construction service provider. Due to the complexity of large construction projects, the standard forms of contract is widely applied. In some cases, although the bespoke construction contract has been adopted, those contracts are often cross-reference clauses in the standard forms of contract.

This chapter introduces the well-used standard forms of contract in large international construction projects. It also provides comparison of the implementation of these well-known standard forms of contract in the civil law and common law jurisdictions and gives recommendations of contract form selection at the beginning of a large capital project.

2.2 STANDARD FORMS OF CONTRACT

Due to the complexity of large construction projects, the use of standard forms of contract provides certain confidence of contract terms to the contract parties. In addition, use of standard forms of contract also saves legal fees during contract drafting and also reduces contract negotiation time which then speeds up the contract agreement.

There are several standard forms of contract used in construction projects, including ICE engineering contracts, FIDIC forms of contract, NEC forms of contract, JCT forms of contract, the PPC2000 contract, the FAC-1 contract, the CIOB Time and Cost Management Contract, the GC/Works contract, the RIBA contract, the RICS contract, ICC conditions of contract, IChemE forms of Engineering contract, IMechE/IET model forms of contract, and the NFDT contract.

In the common law countries, the standard forms of contract is rooted in the private sector. The architects and civil engineers from the Royal Institute of British Architects (RIBA) and Institute of Civil Engineering (ICE) created the JCT and ICE contract, which then developed into the FIDIC and NEC contract. In the United States, the commonly used standard forms of contract is the AIA (American Institute of Architects) contract, which is also developed by architectural practitioners.

In the civil law countries such as France and Germany, the standard forms of contract are rooted in the public sector, where the code of law was created. For example, Vergabe- und Vertragsordnung für Bauleistungen (VOB) is the standard contract for public construction projects in Germany, although it is also widely used in private construction projects.

The use of standard forms of contract is not mandatory in common law jurisdictions, although the standard forms of contract is used as a tradition. In the UK public sector, the GC Works/1 suite of contracts were commonly used and are still available, but they have been largely replaced by the NEC contract, and are no longer being updated.

The FIDIC forms of contract is the most well-known international construction contract. Other well-known standard forms of contract, including the NEC (New Engineering Contract), the JCT (Joint Contracts Tribunal), and the ICC/ICE contract, also play key roles in the construction industry. The annual NBS Construction Contracts and Law Survey (2018) investigated the application trend of each type of construction contract in the construction industry of the United Kingdom. The results illustrate that the most used standard forms of contract in the UK is the JCT contract, which constitutes 62% of usage in 2018. However, although JCT is the most widely used contract form in construction projects in the UK, most large capital projects in the UK use FIDIC and NEC instead. In accordance with the NBS National Construction Contracts and Law Report 2015, for projects of a value between £5m and £25m, NEC is the most used contract with 39% of usage compared to 23% for FIDIC and 18% for JCT; and for projects of a value over £25m, FIDIC takes a substantive lead with 47% of usage, compared to 10% for NEC and 6% for JCT.

Furthermore, there is a trend of increasing use of the NEC contract in the UK for large construction projects and most of the large infrastructure projects are managed under the NEC contract, for example London Crossrail, High Speed 2 (HS2) rail project, 2012 London Olympics associated projects, Highway England framework projects, London Thames Tideway tunnel, environment agency frameworks, and London Heathrow Airport projects. After achieving success in the UK, the NEC contract has gradually been used in public projects in other parts of the world and has achieved extensive success outside of the UK. For example, following the successful trial of 30 public projects in Hong Kong since 2006, the Hong Kong government decided to use NEC as the standard forms of contract for its £7 billion a year public-sector works programme in 2015. Likewise, the South African Construction Industry Development Board also recommends NEC for public sector projects. Furthermore, NEC is growing worldwide in places such as New Zealand, Australia, the Philippines, the Netherlands, Ireland, and China.

Consequently, this book will focus on the discussion of FIDIC, NEC, and JCT forms of contract and associated legal principles of construction law.

2.3 FIDIC

The FIDIC stands for Fédération Internationale des Ingénieurs-Conseils, meaning International Federation of Consulting Engineers. The organization was founded in 1913 by Belgium, France, and Switzerland.

2.3.1 FIDIC Forms

FIDIC published its first standard forms of contract for Civil Engineering works in 1957. The 1957 Red Book is largely based on the fourth edition of ICE (Institute of Civil Engineering) Condition of Contract from the United Kingdom. The original FIDIC forms of contract constitute the 1957 Red Book, the 1963 Yellow Book, and the 1995 Orange Book. The well-used Red Book was published as a fourth edition in

1987. In 1999, FIDIC re-wrote its forms of contract and published a new suite of standard forms of contract including the Green Book for minor work and three forms for main works, namely the 1999 Red Book for Civil Engineering Construction, the 1999 Yellow Book for Design & Build, and the 1999 Silver Book for EPC/Turnkey Project. In addition, FIDIC also published several specific forms. For instance, the White Book is designed for professional consultants or Joint Venture agreements; the Pink Book is published for projects funded by Multilateral Development Banks (MDB), such as the World Bank, the African Development Bank, and the European Bank for Reconstruction and Development (EBRD); the Blue Book is designed for Dredging and Reclamation Works; the Gold Book is designed for DBO (Design, Build, and Operate) projects; the Purple Book is designed for professional consultants; and Sub-Contract 2011 for Sub-contractor Construction for Building and Engineering Works Designed by the Employer. Table 2.1 summaries the FIDIC forms contract to date.

The FIDIC 1999 suites of contracts achieved significant success in managing large construction projects worldwide. Most large international construction projects use FIDIC forms of contract or drafted based on the FIDIC contract. In 2017, 18 years after publishing the first edition of the FIDIC suite, the second edition of the FIDIC rainbow suite of contracts was published, including the Construction Contract second edition (2017 Red Book), Plant and Design-Build Contract second edition (2017 Yellow Book), and EPC/Turnkey Contract second edition (2017 Silver Book). Unlike the 1999 suite, which was a re-write of the previous version of contract, the 2017 suite is not a re-write of the 1999 suite of contracts but provides more explanation and improvement on the first edition of the FIDIC suite, although it contains 50% more text. In 2019, FIDIC published a new form Conditions of Contract for Underground Works (Emerald Book). The severity of financial risks between the Employer and the Contractor of each of the main options is illustrated in Figure 2.1.

Since first published, FIDIC has been well-recognized and applied in the international construction projects. Figure 2.2 explains how to choose the FIDIC forms of contract for specific construction projects. The most well-used forms are the Red Book and the Yellow Book. The Employer is responsible for the design and the Contractor is responsible for the construction in accordance with the Employer's design under the Red Book. For design and build projects, the Yellow Book is usually selected, unless the Contractor needs to provide maintenance of the work, then the Silver Book will be used.

2.3.2 FIDIC Contract Components

The standard FIDIC forms of contract constitutes a condition of contract and supplement document as shown in Figure 2.3.

FIDIC condition of contract is made up of the General Conditions and Particular Conditions. The Particular Conditions take priority over the General Conditions when conflict arises.

2.3.2.1 General Conditions

The General Conditions contain 21 core clauses and the general condition of dispute avoidance/adjudication agreement and DAAB procedure rules. These core clauses

TABLE 2.1
FIDIC Forms of Contract

Forms	Suitable Works	Original	Latest Edition	1999 Suite	2017 Suite
Red Book	Civil Engineering Construction designed by the Employer	1957	2017	1st Edition	2nd Edition
Yellow Book	Electrical and Mechanical works designed by the Contractor	1963	2017	1st Edition	2nd Edition
Orange Book	Design & Build or Turnkey	1995	1995		
Silver Book	EPC/Turnkey Projects	1999	2017	1st Edition	2nd Edition
Green Book	Short Form of Contract	1999	1999	1st Edition	
White Book	Consultant Model Agreements, for appointing a professional consultant – Model Joint Venture (Consortium) Agreement (New in 2017 Edition)	1998	2017	1st Edition	2nd Edition
Pink Book	Construction for Building and Engineering Works Designed by the Employer, and funded by Multilateral Development Banks (MDB)	2005	2010		
Blue Book	Dredging and Reclamation Works	2006	2016		
Gold Book	DBO (Design, Build and Operate) projects	2008	2008		
Sub-contract	Sub-contractor Construction for Building and Engineering Works Designed by the Employer	2011	2011		
Purple Book	FIDIC Model Representative Agreement used for professional consultant	2013	2013		
Emerald Book	Conditions of Contract for Underground Works	2019	2019		1st Edition

apply to most contract forms and are generally consistent in the Red, Yellow, and Silver Books.

Table 2.2 provides the structure of the FIDIC 2017 contract 21 core clauses in nine main categories. Clause 1 provides the general definition for contract terminologies. Unlike NEC4 which only provides 20 defined terms, which is relatively easy to remember. The FIDIC Red and Yellow Books provide 88 and 90 defined terms respectively. Clauses 2 to 4 provide the responsibility of the Employer, the Engineer, and the Contractor respectively. Clauses 5 to 7 describe different resources types, including the sub-contractor, staff and labor, plant, material, and workmanship. Clauses 8 to 10 deal with time management. Clause 11 concerns defects after taking over. Clauses 12 to 14 deal with cost management. Clauses 15 to 16 cover termination. Clauses 17 to 19 handle risk and insurance. Finally, Clauses 20 and 21 provide dispute resolutions.

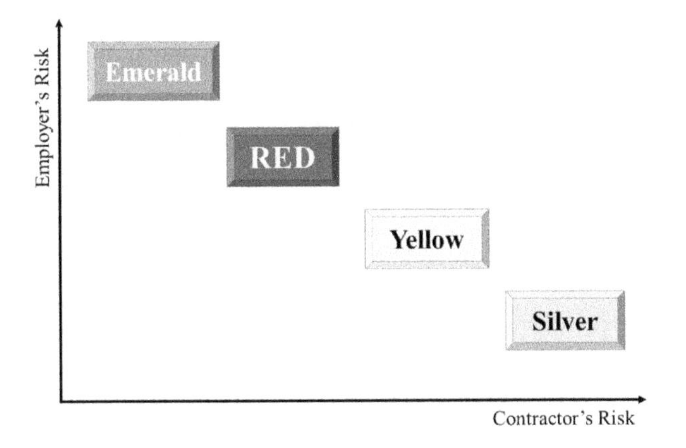

FIGURE 2.1 Financial risk under FIDIC forms of contract.

2.3.2.2 Particular Conditions

Particular Conditions set out the specific conditions for the proposed elements for each project. It provides additional conditions to the General Conditions. Particular Conditions include three components: Part A: Contract Data, Part B: Special Provisions, and building information system.

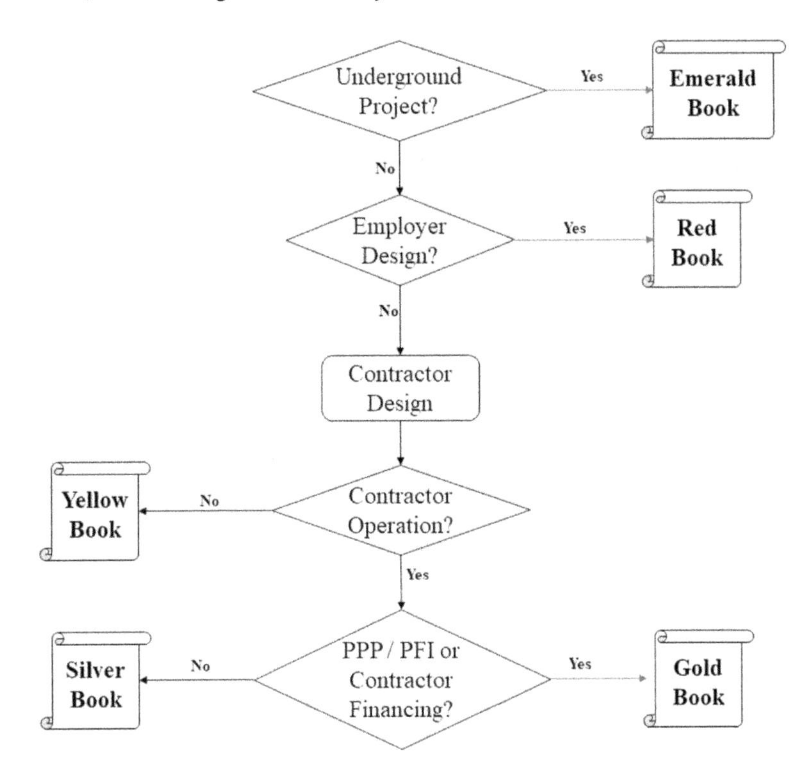

FIGURE 2.2 Selection of FIDIC forms of contract.

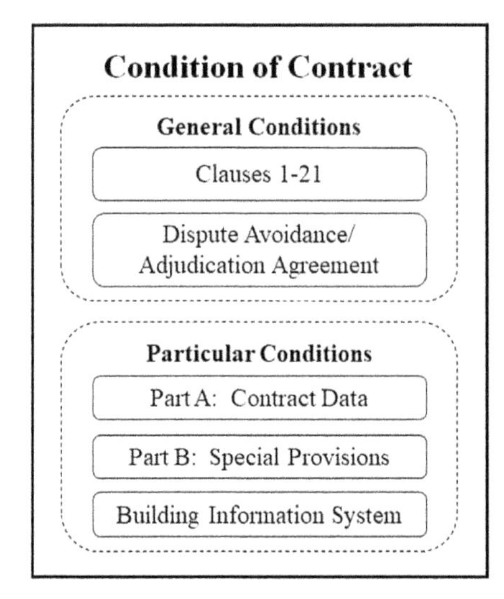

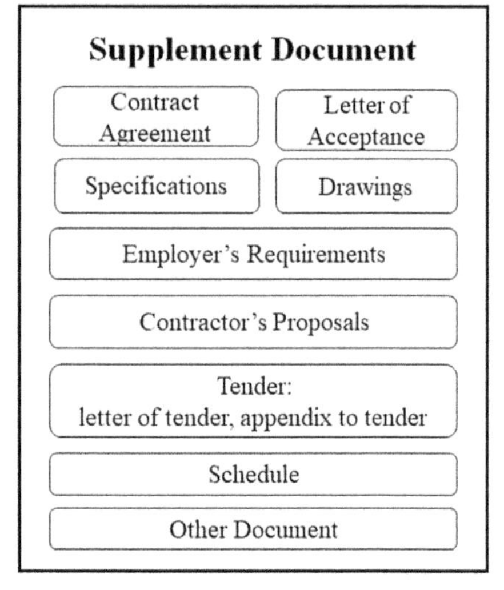

FIGURE 2.3 Content of FIDIC contract.

Contract Data specifies the detailed agreement for each project, for example the Parties' names and addresses, the Time for Completion, the Defects Notification Period (DNP), the communication time limit, the interest rate, the language, the rate of delay damages, insurance requirements, payment period, interim payment, currency, rate of exchange, number of DAAB members and appointment, etc.

Special Provisions work as the Z clause in the NEC contract. It sets out specific works the contract is required to perform or special agreement between the Parties for certain situations. FIDIC recommends the parties limit the modification of the contract conditions and to consider the five golden principles when drafting the Special Provisions as follows:

GP1: *The duties, rights, obligations, roles and responsibilities of all the Contract Participants must be generally as implied in the General Conditions, and appropriate to the requirements of the project.*

GP2: *The Particular Conditions must be drafted clearly and unambiguously.*

GP3: *The Particular Conditions must not change the balance of risk/reward allocation provided for in the General Conditions.*

GP4: *All time periods specified in the Contract for Contract Participants to perform their obligations must be of reasonable duration.*

GP5: *All formal disputes must be referred to a Dispute Avoidance/Adjudication Board (or a Dispute Adjudication Board, if applicable) for a provisionally binding decision as a condition precedent to arbitration.*

The parties are also recommended to use professional legal bodies when drafting the Special Provisions, in order to avoid any conflict with the core clauses of the contract.

With the increasing design digitalization and develop of the building information modeling/building information management (BIM), the specific conditions in regard to the building information system are also set out in the contract. These conditions

TABLE 2.2
Core Clauses under General Conditions of FIDIC 2017

No.	Clause Title
Definition	
1	General Provisions
Parties' Responsibility	
2	The Employer
3	The Engineer
4	The Contractor
Resource	
5	Sub-contracting
6	Staff and Labor
7	Plant, Materials and Workmanship
Time	
8	Commencement, Delays and Suspension
9	Tests on Completion
10	Employer's Taking Over
Quality	
11	Defects after Taking Over
Cost	
12	Measurement and Valuation
13	Variations and Adjustments
14	Contract Price and Payment
Termination	
15	Termination by Employer
16	Suspension and Termination by Contractor
Risk & Insurance	
17	Care of the Works and Indemnities
18	Exceptional Events
19	Insurance
Dispute Resolution	
20	Employer's and Contractor's Claims
21	Disputes and Arbitration

usually include the general standards and procedures, the software/tools to be used, the roles and responsibilities, the data management procedure, the design liability, as well as the copyrights of the design undertaken in the system, etc.

2.3.2.3 Supplement Documents

In addition to the contract provisions, the supplement documents form part of the contract, and are closely linked to the contract conditions. In accordance with Sub-Clause 1.1.10 of the FIDIC 2017 Red Book, the contract constitutes the following documents:

- the Contract Agreement
- the Letter of Acceptance

- the Letter of Tender
- any addenda referred to in the Contract Agreement
- the Conditions of Contract
- the Specification
- the Drawings
- the Schedules
- the Contractor's Proposal
- the JV Undertaking (if applicable)
- further documents (if any) which are listed in the Contract Agreement or in the Letter of Acceptance

Apart from the Contract Agreement and the Conditions of Contract, the remaining elements are part of the Supplement Documents which form the overall contract.

2.4 NEC

Unlike the Joint Tribunal Contract (JCT) which is drafted by lawyers, the New Engineering Contract (NEC) was drafted by Dr Martin Barnes, who was a project manager and an engineer. NEC was developed by Institute of Civil Engineering (ICE). The aim of the NEC contract is to improve management of the construction project. In particular, it encourages a collaborative approach and sets out a clear procedure for programme management and risk management. From the beginning, the NEC contract requests the contract Parties, the *Project Manager* and the *Supervisor*, to work in "*a spirit of mutual trust and co-operation*" in NEC3 (2013) and Clause 10.2 of NEC4 (2017). It aims to resolve problems within the construction project promptly, rather than leaving them to the end of the project as most JCT contracts do. The first edition of the NEC contract was published in 1993, as a result of the ICE's Legal Affairs Committee's suggestion back in 1985.

In order to improve the performance of the construction industry, the UK government instructed Sir Michael Latham to undertake an investigation for the construction industry. In 1994, the Latham Report "Constructing the Team" was published. This report not only investigated the problems in the construction industry, but also provided many valuable recommendations which drove several changes in the UK's construction industry. The most important impact was the implementation of the Housing Grants Construction and Regeneration Act (HGCRA) in 1996, also known as the Construction Act. This Act introduced statutory adjudication for the construction industry and set out a clear payment schedule to improve the cash flow for the contractor, sub-contractor, and supplier. In addition, it recommended both private and public sector use of the NEC contract in engineering and construction projects. In 1995, the second edition of the NEC suite of contracts was published, and the first edition of the NEC contract was renamed to the ECC (Engineering Construction Contract). The most well-used NEC contract is the third edition, which was published in 2005 and was then revised in 2013. Because the UK Office of Government Commerce has endorsed that public projects use the NEC contract since the early 2000s, it became more popular in large infrastructure projects. Almost all the major capital projects in the UK are implemented under NEC forms of contract, for instance the 2012 London Olympics, London Crossrail, London Heathrow Queen Elizabeth Terminal, High Speed 2, Thames Tideway tunnel, etc. In 2017, together with other main standard forms of contract which were updated with new editions, the fourth edition of the NEC suite of contracts was published. Unlike FIDIC 2017 suite of contracts which only updated the Red, Yellow, and Silver Books, NEC4 updated the whole suites of contracts. In addition, NEC4 published two new forms of contract, DBO and ALC. The Design Build Operate Contract (DBO) is used for the same purpose as the FIDIC Gold Book. In order to improve the collaboration as well as Early Contractor Involvement (ECI), NEC4 also published the new Alliance Contract (ALC).

2.4.1 Contract Structure

The NEC contract includes nine sections of core clauses, six main options, three dispute resolution options, 18 secondary options, schedule of cost components, and contract data. Large construction projects undertaken under the NEC contract are usually implemented under the ECC contract. Therefore, this book will focus on the ECC contract when discussing key features of the NEC contract.

The NEC family constitutes a serial form to suit different requirements. There are four main forms of contract including the Engineering and Construction Contract (ECC), the Professional Services Contract (PSC), the Term Service Contract (TSC), and the Supply Contract (SC). Each of these contracts is then derived into relevant short contracts and sub-contracts (excluding Supply contracts). In addition, the Adjudication Contract (AC), which is then replaced by the Dispute Resolution Service Contract (DRSC) are aimed in particular at dispute resolutions. Likewise, the Framework Contract (FC) aims to engage multiple suppliers simultaneously. Further, NEC4 developed the Design Build and Operate Contract (DBO) and Alliance Contract (ALC). Table 2.3 summaries the contract forms which are available in each edition of the NEC contract.

In practice, the typical NEC condition of contract constitutes nine core clauses, plus one main option clause, one dispute resolution option, any secondary options and Schedule of Cost Components (SCC). In addition to the condition of contract, several supporting documents also constitute part of the contract, including Contract Data part one and part two, Price, Scope, and site information. Figure 2.4 illustrates the components of the NEC contract.

2.4.2 Core Clauses

The fundamental principles of the NEC contract are set out in core clauses under nine main sections of the contract, as follows:

1) General
2) The Contractor's main responsibilities
3) Time
4) Quality management
5) Payment

6) Compensation events
7) Title
8) Liabilities and insurance
9) Termination

Each core clause concerns each category, which is also set out in the FIDIC contract. However, the NEC contract is more concise than FIDIC, as the first 23 pages of the contract cover the content of these nine core clauses.

TABLE 2.3
NEC Forms of Contract

Forms	NEC1	NEC2	NEC3	NEC4
Engineering and Construction Contract (ECC)	√	√	√	√
Engineering and Construction Short Contract (ECSC)		√	√	√
Engineering and Construction Sub-contract (ECS)		√	√	√
Engineering and Construction Short Sub-contract (ECSS)		√	√	√
Term Service Contract (TSC)			√	√
Term Service Short Contract (TSSC)			√	√
Professional Service Contract (PSC)		√	√	√
Professional Service Short Contract (PSSC)		√	√	√
Supply Contract (SC)			√	√
Supply Short Contract (SSC)			√	√
Framework Contract (FC)			√	√
Dispute Resolution Service Contract (DRSC)		Adjudicator's Contract		√
Design Build Operate Contract (DBO)				√
Alliance Contract (ALC)				√

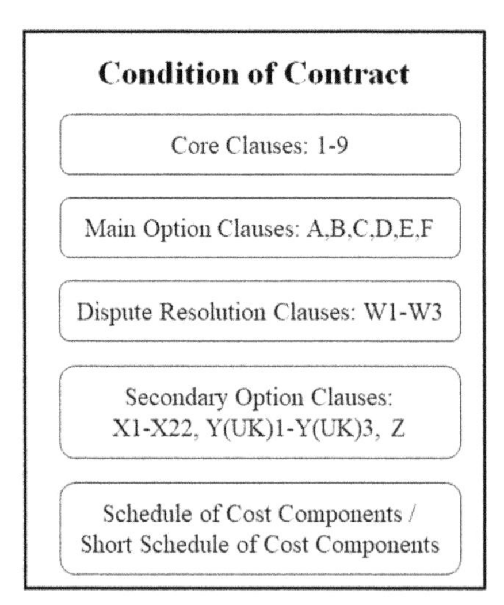

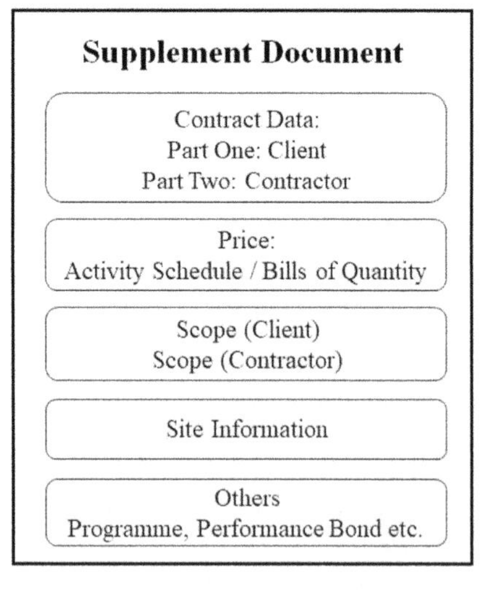

FIGURE 2.4 Components of the NEC contract.

Section 1 sets out the general requirements of the Parties to perform under the contract and act in a spirit of mutual trust and co-operation. Clause 11.1 sets out the identified term under the Contract Data and Clause 11.2 provides definitions of 20 defined terms. Clause 12 concerns the interpretation and the law. Clause 13 sets out the communication requirement. Communication is an important part of the NEC contract. The contract generally requires the Parties to communicate in writing, communicate within the time limit set out in the contract, and to separate each communication. The communication under the NEC contract follows a certain procedure. In practice, large construction projects usually use CEMAR, a contract management tool to facilitate the contract Parties to communicate effectively in accordance with the contract requirements. Clause 14 sets out the duty and authority of the *Project Manager* and the *Supervisor.* In particular, Clause 14.1 confirms that the acceptance of the *Project Manager* or the *Supervisor* does not change the *Contractor*'s obligation under the contract. Clause 15 Early warning is one of the success factors of the NEC contract. In NEC4, the new Clause 16 has been introduced to deal with the *Contractor*'s proposals to echo the changes in Clause 65 of the assessment of the *Contractor*'s proposals. Clause 17 deals with the ambiguity or inconsistency of the Scope. Clause 18 concerns the Corrupt Act. Finally, Clause 19 deals with circumstances under the prevention events, which is similar to the force majeure under the FIDIC 1999 suite of contracts.

Section 2 sets out the *Contractor*'s main responsibilities, including the responsibility of the *Contractor*'s design and using the *Contractor*'s design, the design equipment, the *Contractor*'s key person, working with the *Client* and Others, Subcontracting, Assignment, disclosure, and other responsibilities. The primary obligation of the *Contractor* is how the *Contractor* Provides the Works in accordance with the Scope under Clause 20.1.

Section 3 "Time" is another success factor of NEC in managing large construction projects. It sets out the detailed requirements for the *Contractor*'s programme and the specific procedure to manage the different versions of programme. Chapter 3, Time Management, will explain these features in further detail. In addition, this section also concerns other time related elements, e.g., the stop of work, take over, and acceleration.

Section 4 concerns quality management, which is the duty of the *Supervisor.* In particular it deals with defects of the work. Unlike FIDIC, the Engineer's duties cover managing all time, cost, and quality. In the NEC contract, the *Project Manager* is only responsible for time and cost management, whereas the *Supervisor* is responsible for quality management. As quality management is not within the scope of this book, this section is not discussed in detail in this book. In brief, the Contract Data sets out the specific *defects date* and *defect correction period* for the project. The *Supervisor* can notify Defect up to the *defects date.* The *Contractor* is liable to correct all Defects before the *defects date* regardless of the *Supervisor*'s notification. However, if the *Supervisor* notifies of a Defect, the *Contractor* needs to rectify the Defect within the *defect correction period.* In the event of the *Project Manager* accepting the Defect or appointing Others to correct the Ddefect, the Price will be deducted accordingly, and the Scope is amended.

Section 5 deals with payment, which will be discussed further in Chapter 4, Cost Management. The Construction Act 1996 sets out the statutory obligations of payment responsibility and procedures. Thus, this section is often used in conjunction with secondary option Y1, or Y2. In addition, as different contract main options relate to different payment mechanisms, this section also needs to be used in conjunction with the relevant main option clauses in particular for the definition of the Defined Cost, the Disallowed cost, and the Price for Work Done to Date (PWDD).

Section 6 Compensation Event is another special feature of the NEC form of contract, which will be discussed further in Chapter 6, Change Management. It sets out the 21 entitlements for the *Contractor* to claim additional time and cost under Clause 60.1. It then lays out the specific procedure to notify, quote, assess, and implement the compensation event. NEC4 introduces the new Clause 65 for proposed instructions.

Section 7 concerns the title of the work and materials on site, which is not within the scope of this book.

Section 8 concerns liability and insurance. NEC4 changes the term "risk" in NEC3 to "liability." Chapter 5, Risk Management, will discuss the related clauses further.

Finally, Section 9 deals with termination, which is also not within the scope of this book.

2.4.3 Main Options

The Parties using the NEC contract need to choose one of the main options depending on the way they intend to manage the contract. The NEC contract provides six main options, as follows:

- Option A: Priced contract with activity schedule
- Option B: Priced contract with bill of quantities
- Option C: Target contract with activity schedule
- Option D: Target contract with bill of quantities
- Option E: Cost reimbursable contract
- Option F: Management contract

Option A and Option B are fixed price contracts. The *Contractor* is paid based on the completed Activity Schedule under Option A, whereas it is paid on fixed Bills of Quantity (BoQ) under Option B. Option C and Option D are target price contracts with the pain/gain mechanism. Likewise, Option C is based on the activity schedule and Option D is based on Bills of Quantity. Option E is a cost reimbursable contract, whereas Option F is a management contract. The risk severity between the *Client* and the *Contractor* of each of the main options is illustrated in Figure 2.5.

In practice, for the large infrastructure projects undertaken in the UK, Option C is often used to form the contract between the *Client* and the main *Contractor*, and Option A is often used for the contract between the main *Contractor* and its subcontractors, or for the contract between the *Client* and the management *Consultant*. For projects with less uncertainty of Scope, Option E is often used.

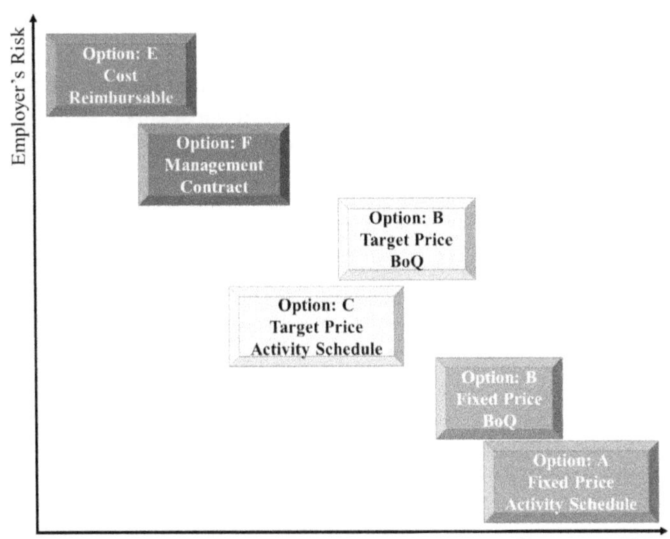

FIGURE 2.5 Financial risk associated to the main options of NEC forms of contract.

2.4.4 Secondary Options

The secondary option clauses of the NEC4 contract include:

Option X1: Price adjustment for inflation (used only with Options A, B, C, and D)
Option X2: Changes in the law
Option X3: Multiple currencies (used only with Options A and B)
Option X4: Ultimate holding company guarantee
Option X5: Sectional Completion
Option X6: Bonus for early Completion
Option X7: Delay damages
Option X8: Undertakings to the Client or Others
Option X9: Transfer of rights
Option X10: Information modeling
Option X11: Termination by the Client
Option X12: Multiparty collaboration (not used with Option X20)
Option X13: Performance bond
Option X14: Advanced payment to the Contractor
Option X15: The Contractor's design
Option X16: Retention (not used with Option F)
Option X17: Low performance damages
Option X18: Limitation of liability
Option X19: Termination by either party (Terms Service Contract only)
Option X20: Key Performance Indicators (not used with Option X12)
Option X21: Whole life cost

Option X22: Early Contractor involvement (used only with Options C and E)
Option Y(UK)1: Project Bank Account
Option Y(UK)2: The Housing Grants, Construction and Regeneration Act 1996
Option Y(UK)3: The Contracts (Rights of Third Parties) Act 1999
Option Z: Additional conditions of contract

The Z clause is often used to set out additional conditions of the contract. However, practice experience suggests that the use of such clauses can create comprehensive issues during the contract implementation. Thus, it is recommended to limit the use of this clause and the Parties should engage with competent legal and NEC contract professionals to draft these conditions.

2.5 JCT

The Joint Contracts Tribunal (JCT) contract was created in 1931 by the Royal Institute of British Architects (RIBA). JCT was joined by the RICS contract in 1947, by local authorities' and specialist sub-contractors' associations in 1963, and by the Association of Consultant Architects (ACE) only from 1975 to 1976. Following the issuing of the first standard form of JCT contract in 1931, later versions were published in 1937, 1939, 1963, 1980, 1998, 2011, and 2016. In accordance with the NBC's national construction survey in the UK, JCT is the most used standard form of contract in the UK. It is mainly used in projects relating to residential and commercial buildings, which are of low to medium value. Although less used in large construction projects, which is the focus of this book, as the most popular forms of contract in the UK construction industry, it deals with all the key legal issues of the construction project. Therefore, when drafting something bespoke, the provisions under the JCT contract may also be considered.

2.5.1 JCT Forms

In 2016, JCT published the new suite of contract forms and agreements, including 13 types of contract forms and five agreements across a wide range of contract and sub-contract forms as follows:

Contract forms:

- Standard Building Contracts (SBC)
- Intermediate Building Contracts (IC)
- Minor Works Contracts (MW)
- Major Project Contracts (MP)
- Design & Build Contracts (DB)
- Management Building Contracts (MC)
- Construction Management Contracts (CM)
- Constructing Excellence Contracts (CE)
- Measured Term Contracts (MTC)
- Prime Cost Building Contracts (PCC)
- Repair & Maintenance Contracts (RM)
- Home Owner Contracts (HO)
- Short Form of Sub-Contract (ShortSub)

Agreements:

- Framework Agreements (FA)
- Collateral Warranties (WCWa)
- Pre-Construction Services Agreements (PCSA)
- Adjudication Agreements (Adj)
- Consultancy Agreements (CA)

This broad range of contract forms provides the opportunity for the parties to choose the most appropriate form in accordance with the project features and risk appetite of the Employer.

2.5.2 JCT Contract Components

The JCT contracts start with the agreement between the Contractor and the Employer's recitals, which is similar to the details set out in the Contract Data under the FIDIC and NEC contracts. It then compromises nine articles, contract particulars, attestation, nine conditions, and seven schedules.

The JCT contract sets out the Parties agreement at the beginning of the contract, the nine articles set out the specific general Contractor's obligations, contract parties, value, and dispute resolution and legal proceedings as follows:

1) Contractor's obligations
2) Contract Sum
3) Employer's Agent
4) Employer's Requirements and Contractor's Proposals
5) Principal Designer
6) Principal Contractor
7) Adjudication
8) Arbitration
9) Legal proceedings

The contract conditions are set out in nine sections with the general obligation, key personnel, and dispute resolution and legal proceedings. It is the core of the contract, which includes the following:

Section 1 Definitions and Interpretation
Section 2 Carrying out the Works
Section 3 Control of the Works
Section 4 Payment
Section 5 Changes
Section 6 Injury, Damage and Insurance
Section 7 Assignment, Performance Bonds and Guarantees, Third Party Rights and Collateral Warranties
Section 8 Termination
Section 9 Settlement of Disputes

Following the General Conditions, the optional provisions are then provided in seven schedules as follows:

Schedule 1 Design Submission Procedure
Schedule 2 Supplemental Provisions
Schedule 3 Insurance Options
Schedule 4 Code of Practice
Schedule 5 Third Party Rights
Schedule 6 Forms of Bonds
Schedule 7 JCT Fluctuations Option A

2.6 COMMON LAW

English law plays an important role in the construction contract because the most widely used standard forms of contract are sourced from the English contract. FIDIC was based on the fourth edition of the ICE contract. NEC and JCT are both developed in the UK.

The essential basis of English common law is case law, which is established on the doctrine of precedent. It focuses on detailed analysis of the material facts of a particular situation leading to the development of precise rules applicable to the given situations. With the notion of freedom of contract, parties are flexible to enter into a contract and draft provisions for specific types of project.

2.6.1 Contract Formation

Under English contract law, the incorporation of a valid binding contract requires meeting three essential elements, which are the intention to create legal relationship, agreement, and consideration. There is a presumption that there is an intention to create a legal relationship between the contract parties' in commercial agreements. This presumption can only be rebutted by express terms. The agreement is established through offer and acceptance. Consideration is the essential requirement for a binding contract. The case law developed the rules of consideration and the valid consideration must move from the promise, must be present, and must be sufficient.

2.6.2 Contract Interpretation

Over the decades, the court also developed the rules to interpret contract terms in the event of dispute on the contract terms. Under English law, the basic principles of contract interpretation were laid down by Lord Hoffman in *Investors Compensation Scheme Ltd. v West Bromwich Building Society* [1998] 1 WLR 896. It is also known as the "Canons of construction" which includes:

1) objective intention of the parties
2) matrix of facts
3) exclusions of pre-contract negotiations and subjective intent
4) the meaningful is not a matter of dictionaries but the whole contract
5) detailed syntactical analysis yields to business common sense

The express provision in the contract is the primary source for interpreting the contract. The parties' intention is assessed objectively based on the standing point of a reasonable person having all background knowledge that was available to the parties in the situation. To avoid pedantic or literal interpretation, the court will consider business common sense in contract interpretation, but the *contra proferentem* rule may apply. For example, Clause 63.10 of the NEC contract provides that the compensation event arising from the ambiguity or inconsistency of the Scope will be interpreted in favor of the Party not providing the Scope.

2.6.3 Good Faith

Under the common law, if the contract provision expressly sets out the obligations of good faith, then the parties have the duty to act in good faith. Under English law, the court is reluctance to adopt the principle of implied good faith. However, some piecemeal solutions have been developed to resolve the unfairness in the English case law.

Nevertheless, some courts in the United States recognize the implied good faith in the commercial contract. Some common law jurisdictions also accept implied duty of good faith.

2.6.4 Defect Liability

English construction contracts usually contain a defined defects liability period after completion. For large construction projects, the defects liability period is generally between one to two years. The Contractor is liable to rectify all defects during this period, regardless notified to him or not. The Contractor's liability after defects liability period will be imposed by the Limitation Act 1980. The Employer can claim damages within six years from breach of the contract or 12 years from breach of the deed, although the Parties may agree to a shorter limitation period separately.

2.6.5 Damages

In common law jurisdiction, the damages of economic loss for breach of contract is subject to the "remoteness" test, which is set out in the leading case *Hadley v Baxendale* (1854) 9 Ex 341. The court sets out two limbs for the "remoteness" test considering the direct loss and indirect loss respectively. The first limb considers the direct loss arising naturally, according to the usual course of things, from the breach of contract with common knowledge. The second limb considers indirect loss which does not arise naturally but reasonably within the contemplation of both parties at the time the contract was incorporated as the probable result of the breach, therefore requiring special knowledge. The Parties are liable for direct and indirect loss unless contractually excluded in the contract.

2.6.6 Termination

The doctrine of freedom of contract under common law also extends to freedom to restrict liability. Hence, the parties can exclude certain liabilities with mutual agreement, except as regards death, personal injury, and fraud which is governed by the regime of the criminal law.

Under common law, the contract can be terminated by agreement, by performance, by breach of the contract, or by frustration.

The court developed the doctrine of frustration in order to relieve the party from the contract, when external circumstances prevent its performance of contract as originally intended was beyond its control. A contract may be frustrated due to further performance is impossible, illegal, or radically different; limited applicability in practice, leading to frequent contractual recourse to force majeure; or under type provisions.

Repudiation is a special feature of breach of contract under common law. Repudiatory breach is a sufficiently serious breach that allows the innocent party to be discharged from the contract obligation and to terminate the contract, for instance, if the Contractor refuses to perform the work or leaves the site, or the Employer fails to give access to the site, or appoints another Contractor to perform the works in the scope of the contract.

Provisions of termination for convenience is increasingly common in standard forms of contract, though such a provision is of limited benefit to the Contractor. The English courts respect the contract provision that gives a party full discretion to terminate at will.

2.7 CIVIL LAW

In contrast to the common law, civil law is a codified constitution. The first civil law can be traced back to Code of Hammurabi in Babylon in ca. 1790 BC. The civil law system is derived from the Roman law. It then developed into different branches including the French civil law, the German civil law, the Scandinavian civil law, the Chinese civil law, and the Islamic civil law. The standard forms of contract can have different implementations under civil law jurisdictions. As a member of European Union since 1973, EU civil law legislation has impacted the English law as well. This section will use French law as an example of civil law to explain the differences to the common law system.

2.7.1 Sources of Law

French law is derived from the Napoleonic Code. As one of the major civil law systems in the world, Napoleonic civil law countries include France and its overseas territories, Italy, Spain and their former colonies in Latin America, Francophone Africa, the Middle East, and South East Asia.

The primary source of law is the French Civil Code (1804). It is a comprehensive body of legislation drafted in terms of principles, leaving the courts to decide the precise meaning of the text. It sets out the mandatory and non-mandatory provisions. The mandatory provisions cannot be superseded by the contract. The French Civil Code (1804) is then supported by the commercial code and administrative law. Case law can contribute to the source of French law, but it is only used when the law is unclear. However, in contrast to the common law, the decision of the case does not bind the following cases.

2.7.2 Formation of a Contract

Under French law, the general conditions for the validity of a contract are set out in Article 1108 of the French Civil Code as follows:

- the consent of the parties
- the party's legal capacity to enter into the contract
- a specified objective of the agreement
- a lawful cause/valid reason for the agreement

Article 1159 of the French Civil Code provides that in case of ambiguity of the contract, it must be interpreted in accordance with that of the country where it is concluded.

In construction contracts, Article 1710 of the Civil Code provides special requirements of price certainty. Therefore, a detailed pricing system may be specified in the contract and the *quantum meruit* in common law does not exist in the civil law system.

2.7.3 Good Faith

Because the civil law jurisdictions follow the Roman law root, which forms the contract in the notion of the German Civil Code of *bona fides*, good faith is a general duty embedded in all contracts. The EU Directives generally enforce the duty of good faith. For example, the contract parties have the duty of good faith under both French law and German law.

Paragraph 3 of Article 1134 in the French Civil Code requires that all parties to a contract must perform their obligations in good faith as follows:

> *Agreements lawfully entered into take the place of the law for those who have made them. They may be revoked only by mutual consent, or for causes authorized by law They must be performed in good faith.*

Likewise, Article 242 of the German Civil Code (Bürgerliches Gesetzbuch – BGB) provides:

> *An obligor has a duty to perform according to the requirements of good faith, taking customary practice into consideration.*

The good faith obligation also extends to the period where the parties negotiate the contract. Therefore, the consultants are obliged to comply with the well-used standard in the industry. The Contractor must conform to the "state of the art," and the Contractor has the duty to warn the Employer of this, echoing the requirement of the FIDIC contract.

2.7.4 Types of Contract: Public or Private

Public and private projects have different forms of contract under French law. Private projects use the codified contract law, whereas public projects implement un-codified public contract under the administrative law. Consequently, two different set of courts have a distinct jurisdiction up to the highest level. For the private contract law, the highest court is the Cour de cassation; whereas, for the public contract law, the highest court is le Conseil d'État. The public law concerns the general interests, whereas the private law concerns the private interests. As the contracts between the Contractor and the government are not only concluded for the interest of each party, but also for the general interest and the benefit of the public at large, in the case of general interest conflicting with private interest, the public interest overrides the private interest and may justify legal rights and obligations not available in the private contract, as exorbitant rights of common law.

2.7.5 Standard Forms of Contract

Standard fixed price and cost remeasurement contracts are widely used in France as well as in many Napoleonic civil law countries. In France, the standard contract for private works is AFNOR standards conditions, and the standard contract for public works is CCAG standards conditions, which is a cost remeasurement contract. Both private and public contracts have a long industry practice history with well-published commentary and guidelines.

In general, FIDIC is not commonly used in France although use of the FIDIC Silver Book is gradually increasing. For projects undertaken with the World Bank or Development Bank fund in Francophone Africa, the FIDIC Pink Book (MDB) is generally used.

2.7.6 Limitation of Liability

Under French law, the Consultant has reasonable skill and care liability, whereas the Contractor has the strict liability if the Employer proves the damages are the Contractor's fault. The Consultant often shares responsibility of delay with the Contractor and the Consultant may have to indemnify the Contractor when its delay has caused an overall delay of over 25%.

In contrast to English law, which only provides six years limitation liability for contracts, French law provides 30 years contractual liability. Under French law, there is a presumption of the Contractor's liability for works undertaken by it, and the Employer does not need to prove the Contractor is at fault. French law also sets out a one-year defect liability period after the Employer's take over, two years liability for equipment related elements, and ten years liability for civil and structural engineering defects together with mandatory insurance covers.

2.7.7 Liquidated Damages and Penalty

Under English law, the use of penalty clause will waive the rights for liquidation damages. In contract, under French law, Article R231-14 of Code de la construction et de l'habitation expressly sets out that penalties clause for delay and the damages for delay should be no less than 1/3000 of the agreed contract price per day of delay. It also sets out the penalty for late payment which is limited to 1% per month of the outstanding payment.

2.7.8 Termination

The right to terminate is distinct between private and public work contracts under French law. Before the French contract law reform in 2016, both private contracts and public contracts can terminate in event of force majeure. However, the new contract law sets out new rules for terminating the contract. Article 1147 provides:

> *An obligor is liable for damages arising either from non-performance or from delay in performance of the obligations, unless he can show that his failure to perform was caused by events beyond his control and that there was no bad faith on his part.*

In order to give rise to terminate the contract, the event must achieve three conditions: unforeseeable, unavoidable, and external.

For private projects, a party can terminate the contract in the circumstances of breach obligations, and/or lack of payment. In addition, the Employer also has the right to terminate for convenience under a fixed price contract. In such circumstances, the Contractor is entitled to claim payment for losses of profit. Furthermore, the Employer also has the right to terminate when it is justified by public interest, or when the severe delay can prejudice the Employer. Under common law, the party has no general right to suspend, unless it is expressly provided in the contract. However, in private contracts under French law, both the Employer and the Contractor have a general right to suspend performance if the other party is not performing its obligations under the contract. Therefore, each party has the right to terminate the contract, when a material breach of contract arises by the other party.

In public projects, if the Contractor breaches the contract, then the contract terminates automatically from the time of breach. Whereas, if the Employer breaches the contract, only a judge can decide whether the contract can be terminated. In the absence of express provision, insolvency itself does not give rise to termination under English law. French law expressly prohibits clauses providing for termination upon commencement of insolvency under L.622-13 of Code de commerce.

2.8 RECOMMENDATIONS

Well begun is half done. The choice of construction contract is important for the success of large construction projects. There are pros and cons to each of the standard forms of contract. JCT has been used for over 80 years, it has been drafted with thorough consideration of legal principles and provides more legal certainty. However, many problems that occur during the project cannot be resolved in time and often result in a lengthy dispute after completion. NEC was created to improve collaboration of parties engaged in the construction project, using a proactive approach to resolve the problems as the project progress. However, it requires more project management input throughout the project life cycle. FIDIC has received recognition worldwide, but the new 2017 suite of contracts may still be subject to practical tests and may need further amendments to achieve the expected improvements.

In recent years, the framework contract and early contract involvement led to savings in large construction projects. Therefore, both NEC and JCT also provide the framework contract and forms for early contract involvement. FAC-1 is an alliance contract which is increasingly used in public projects and resulted in savings of about 20%.

In addition to the contrast between the contract forms, the risk allocation between the contract parties should also considered carefully. Unbalanced risk allocation often results in the breaking off of relationships between contract parties and lengthy disputes towards the end of the projects. Choosing the contract option with the lowest risk to a party may not be good to the overall project. The author would suggest the parties work collaboratively throughout the project and allocate the risk to the parties most capable to handle it and aim to achieve overall project success for both parties.

REFERENCES

Book/Article

FIDIC. 1999. *Conditions of Contract for Construction.* 1st Edition. (1999 Red Book). Geneva, Switzerland: The Fédération Internationale des Ingénieurs-Conseils.

FIDIC. 2017. *Conditions of Contract for Construction.* 2nd Edition. (2017 Red Book). Geneva, Switzerland: The Fédération Internationale des Ingénieurs-Conseils.

FIDIC. 2017. *Conditions of Contract for EPC / Turnkey Project.* 2nd Edition. (2017 Yellow Book). Geneva, Switzerland: The Fédération Internationale des Ingénieurs-Conseils.

FIDIC. 2017. *Conditions of Contract for Plant & Design Build.* 2nd Edition. (2017 Silver Book). Geneva, Switzerland: The Fédération Internationale des Ingénieurs-Conseils.

Latham, Michael. 1994. *Constructing the Team: Final Report of the Government / Industry Review of Procurement and Contractual Arrangements in the UK Construction Industry.* London, UK: HMSO.

NBS. 2018. *National Construction Contracts and Law Report 2018.* https://www.thenbs.com/knowledge/national-construction-contracts-and-law-report-2018, Accessed on 12 May 2019.

NEC. 2013. *NEC3 Engineering and Construction Contract.* London, UK: Thomas Telford Ltd.

NEC. 2017. *NEC4 Engineering and Construction Contract.* London, UK: Thomas Telford Ltd.

Legislation

French Civil Code 2016.

Bürgerliches Gesetzbuch (German Civil Code). 2002. http://www.ilo.org/dyn/natlex/natlex4.detail?p_lang=&p_isn=61880&p_classification=01.03, Accessed on 10 November 2019.

Code de la construction et de l'habitation. 2019. https://www.legifrance.gouv.fr/affichCode.do?cidTexte=LEGITEXT000006074096, Accessed on 10 November 2019.

Case

Hadley v Baxendale (1854) 9 Ex 341 24.

Investors Compensation Scheme Ltd. v West Bromwich Building Society [1998] 1 WLR 896.

3 Time Management

3.1 INTRODUCTION

Project planning often plays a crucial role in major construction projects. Due to the complexity of construction projects, generally one third of construction projects complete after the initial contract completion date. Due to the scale of large international construction projects, the liquidated damages associated with delay of the project can be significant. Therefore, the legal issues associated with project planning often become a critical aspect in claims and disputes of large international construction projects. This chapter starts with an explanation of the legal obligation of completion of the construction project under both contract express provisions and statutory implied terms. It then discusses the delay claim and associated legal principles, including prevention principle, extension of time, time at large, liquidated damages, concurrent delay, etc. With the lessons learned from existing cases, the best practice of project planning at the front-end of a project is recommended.

3.2 LEGAL OBLIGATION OF COMPLETION

All projects have a completion obligation. The project contract usually sets out a contract completion date. Even if the contract is silent on the date of project completion, the common law and statute can impose completion obligation by implied terms.

Upon completion, the timer calculating the liquidated damages stops (e.g., NEC Option X7). The Contractor is then entitled to a partial retention payment if there is any (e.g., 50% under NEC Option X16); the defects liability period starts and is fixed; the Employer takes over the work, and insurance liability transfers from the Contractor to the Employer.

3.2.1 Terminology of Completion

In practice, different standard forms of contract use different terminology for the completion, such as substantial completion, practical completion, completion, and planned completion. It is worth clarifying the meaning and source of each term in order to use it appropriately in the relevant forms of contract. Table 3.1 summarizes the terminology related to completion in the popularly used standard forms of contract.

3.2.1.1 Substantial Completion

"Substantial Completion" is a concept adopted in the FIDIC contract and ICE forms contract. In the FIDIC 1987 Red Book, Sub-Clause 48.1 provides:

> *When the whole of the Works have been* ***substantially completed*** *and have satisfactorily passed any Tests on Completion prescribed by the Contract....*

TABLE 3.1
Terminology of "Completion" in Standard Forms of Contract

Contract Form	Contractual Deadline	Actual Finish Time
NEC	Completion	Completion Date
FIDIC	Substantial Completion	Time for Completion
JCT	Practical Completion	Date for Completion

In FIDIC 1999 suite of contracts, Sub-Clause 10.1(a) states:

> *issue the Taking-Over Certificate to the Contractor, ...except for any minor outstanding work and defects which will not* ***substantially*** *affect the use of the Works or Section for their intended purpose....*

Likewise, in the FIDIC 2017 suite of contracts, Sub-Clause 10.1(i) provides:

> *issue the Taking-Over Certificate to the Contractor, ...except for any minor outstanding work and defects...which will not* ***substantially*** *affect the safe use of the Works or Section for their intended purpose....*

3.2.1.2 Practical Completion

"Practical Completion" is a concept under the JCT contract; many building contracts in the UK also refer to "Practical Completion." However, the JCT contract in general does not provide a clear definition of "Practical Completion." The definition is only clear in the JCT Major Project Construction Contract, 2016 edition, as follows:

> ***Practical Completion*** *takes place when the Project is complete for all practical purposes and, in particular:*
>
> - *the relevant Statutory Requirements have been complied with and any necessary consents or approvals obtained;*
> - *neither the existence nor the execution of any minor outstanding works would affect its use;*
> - *any stipulations identified by the Requirements as being essential for Practical Completion to take place have been satisfied; and*
> - *the health and safety file and all 'As-Built' information and operating and maintenance information required by this Contract to be delivered at Practical Completion has been so delivered to the Employer.*

In the absence of a clear definition of "Practical Completion" in the contract, the contract interpretation principles under common law cases may apply. In *Mariner International Hotels Ltd v Atlas Ltd (2007)* 10 HKCFAR 1, the court recognized practical completion as follows:

> *as used in building contract '****practical completion****' is a legal term of art well understood to mean a stage of affairs in which the works have been completed free from patent defects other than ones to be ignored as trifling.*

3.2.1.3 Completion

The NEC and PPC2000 contracts simply refer to "Completion." Clause 11.2(2) of the NEC4 contract clearly defines that:

> ***Completion*** *is when the Contractor has done all the work which the Works Information states he is to do by the Completion Date and corrected notified Defects which would have prevented the Employer from using the works and Others from doing their work.*

3.2.1.4 Planned Completion

The NEC contract specifically requires the Contractor to show the "planned Completion" in the programme separately from the "Completion Date" of the contract under Clause 31.2. It is important to understand the difference between these two terms. The contract Completion Date can only be changed under three circumstances as follows:

1) extending the time to move the Completion Date to a later date through a compensation event under Clause 60.(1) plus other contract entitlements explained in Chapter 6
2) acceleration to bring forward the completion date under Clause 36
3) changes implemented under Option X21 "Whole Life Cost"

Clause 14.1 clearly provides that change of programme planned Completion date in the Accepted Programme does not change the Contractor's obligation of Completion under Clause 30.1.

3.2.1.5 Completion Date or Time/Date for Completion

The "Completion Date" under NEC, the "Date for Completion" under JCT, or "Time for Completion" under FIDIC is the date that is set out in the contract by which the work is required to be finished. It is the date used to start counting the time of the delay damages; whereas the Completion is the actual date by which the Employer takes over the works, and the date stops the clock for counting delay damages.

3.2.2 Contract Express Provisions

When the completion date is set out in the contract, the Contractor must complete the works by the Completion Date in accordance with the terms of the contract. Most standard forms of contract set out express provision for both Completion Date, which is the deadline to deliver the project, and the Completion, which is the requirement to fulfil the Contractor's obligation on the project.

3.2.2.1 FIDIC

The definition of "Time for Completion" under Sub-Clause 1.1.84 of FIDIC 2017 Red Book refers to the meaning under Sub-Clause 8.2 [Time for Completion] and Sub-Clause 8.5 [Extension of Time for Completion], as follows:

> *the time for completing the Works or a Section (as the case may be) under Sub-Clause 8.2 [Time for Completion], as stated in the Contract Data as may be extended*

under Sub-Clause 8.5 [Extension of Time for Completion], calculated from the Commencement Date.

Sub-Clause 8.2 "**Time for Completion**" then sets out the requirements for completion as follows:

The Contractor shall complete the whole of the Works, and each Section (if any),within the Time for Completion for the Works or Section (as the case may be), including completion of all work which is stated in the Contract as being required for the Works or Section to be considered to be completed for the purposes of taking over under Sub-Clause 10.1 [Taking Over the Works and Sections].

Therefore, the Completion under the FIDIC contract is related to the relevant taking-over requirements.

3.2.2.2 JCT

Clause 1.1 of JCT DB 2016 defines completion date as:

the Date for Completion of the Works or of a Section as stated in the Contract Particulars or such other date as is fixed under Clause 2.25 or by a Pre-agreed Adjustment.

The guidance note of JCT 2016 contract defines the term "**Date for Completion**" as:

The date by which the Contractor is required to finish the work, as stated in the Contract Particulars or subsequently extended by the Architect / Contract Administrator.

Clause 2.25 "Fixing Completion Date" then provides the conditions that the "Date for Completion" can be changed after contract formation, for instance through variations.

3.2.2.3 NEC

Clause 11.2(2) defines **Completion** as when the Contractor has: 1) completed all the work required in the Works Information stated by the Completion Date; and 2) corrected notified Defects which would prevent the Employer using the *works*. Clause 30.1 provides that "*The Contractor...does the work so that Completion is on or before the* ***Completion Date***." Clause 11.2(3) then provides that the Completion Date is the completion date stated in the Contract Data, which is decided by the contract Parties, unless changed later through operating relevant contract clauses, for example, Clause 36.3 acceleration, Clause 65.3 implementing the compensation event, or Clause X21 implementing changes to improve the project's Whole Life Cost.

Unlike FIDIC, the definition of Completion in the NEC contract does not refer to the takeover, because the Client can take over part of the works in advance under Clause 35.3, while the contractual Completion must finish all the works.

3.2.2.4 PPC2000

PPC2000 amended in 2013 (Mosey, 2014) defines the Completion Date in Appendix 1 as "*the date that the Project achieves Project Completion in accordance with Clause 21 of the Partnering Terms.*" Completion of the project in accordance with the Patterning Documents necessary for the Client to use and occupy the Project to the agreed standards.

The completion does not release the Contractor's liability completely from the work; the Contractor still needs to correct any defects occurred during the defect liability period, which is usually one to two years after completion.

3.2.3 Implied Terms

If the contract is silent on a specific completion date, the Contractor still has an implied obligation to complete the project within a reasonable time imposed by statute and/or common law.

3.2.3.1 SGSA 1982

Section 14(1) of the Supply of Goods and Services Act 1982 provides that:

> *Where, under a [relevant contract for the supply of a service] by a supplier acting in the course of a business, the time for the service to be carried out is not fixed by the contract, left to be fixed in a manner agreed by the contract or determined by the course of dealing between the parties, there is an implied term that the supplier will carry out the service within a reasonable time.*

3.2.3.2 Consumer Rights Act 2015

Likewise, Section 52 of the Consumer Rights Act 2015 provides:

> (1) *This section applies to a contract to supply a service, if:*
>   a) *the contract does not expressly fix the time for the service to be performed, and does not say how it is to be fixed, and*
>   b) *information that is to be treated under section 50 as included in the contract does not fix the time either.*
>
> (2) *In that case the contract is to be treated as including a term that the trader must perform the service within a reasonable time.*

3.2.3.3 Common Law

The English common law also imposes the Contractor's implied obligation to complete the works within a reasonable time. In addition, business efficacy may also be imposed to the Contractor for performing with due diligence.

In *Hick v Raymond and Reid* [1893] A.C. 22, the House of Lords held that the Contractor fulfilled his obligation to complete within a reasonable time because the delay caused by striking dockworkers was beyond its control. Lord Watson ruled the assessment of the "reasonable time" as follows:

> *the party upon whom it is incumber duly fulfils his obligations notwithstanding the protracted delay, so long as such delay is attributable to causes beyond his control and he has neither acted negligently nor unreasonably.*

Further, the reasonableness of time is also determined by the relevant circumstances at the time the question arises as addressed by the Court of Appeal in *Shawton Engineering Ltd v DGP International Ltd* [2005] EWCA Civ 1359.

Both the Employer and the Contractor may contribute to the delay in construction projects. Under English common law, if the Employer's action prevents the Contractor from performing or an unrealistic time bar is applied in the variation process, the original contract completion date falls away and the Contractor can then complete the works within a reasonable time. This will be discussed in more detail in the "prevention principle" and "time at large" in the following sections.

3.2.4 Time Is of the Essence

Time being of the essence means that the contract terms of completion date is a condition instead of a warranty or is innominate. Therefore, if time is of the essence for the contractual completion date, the failure to complete the project by contract completion date amounts to a breach of contract and the Employer can treat the contract as repudiated. As a result, the Employer is entitled to terminate the contract and waive its liability to pay for uncompleted works. For example, in *Union Eagle Ltd v Golden Achievement Ltd* [1997] UKPC 5, 10 minutes of delay triggered termination of a property purchase contract and the buyer lost a large deposit.

In construction contracts, time is generally not of the essence. However, time of the essence may apply to some individual terms. The common law position of time of the essence is ruled in *United Scientific Holdings Ltd v Burnley Council* [1978] A.C. 904 as follows:

> *Time will not be considered to be of the essence unless: (1) the parties expressly stipulated that conditions as to time must be strictly complied with; or (2) the nature of the subject matter of the contract or the surrounding circumstances show that time should be considered to be of the essence; or (3) a party who has been subject to unreasonable delay gives notice to the party in default making time of the essence.*

Likewise, paragraph 263 of volume 6 of the building contract of Halsbury's Laws of England (2019) provides that time is not of the essence under the following circumstances:

1) in the absence of express contract provision or necessary implication
2) weekly payments are made for the uncompleted works after contract completion date
3) when parties consider postponing the completion date
4) in the absence of specific completion date or the completion date cannot be precisely decided

In accordance with paragraph 264 of volume 6 of the building contract of Halsbury's Laws of England (2019), even if time is not of the essence, when the completion date is specified in the contract, the Employer can claim damages for the delay that has arisen due to the fault of the Contractor.

Therefore, the contract parties should take careful consideration during the contract drafting stage, as making time of the essence may have a significant impact on the project later on.

3.3 PROJECT PLANNING

In order to complete the works by the contract completion date, the project plan is developed to manage the complex construction activities, monitor progress, control changes, and is used as a basis for making or defending a time claim.

3.3.1 Programme

3.3.1.1 Baseline Programme

The baseline programme is important for all projects, as it establishes the basis where the schedule control and earned value analysis are undertaken. The baseline programme is particularly crucial to control the changes in large international construction projects and it is often used as the benchmark for delay and disruption analysis. In the absence of an appropriate baseline programme, the Employer cannot manage and control the project effectively. The sanction Clause 50.5 for the first programme of the NEC contract ensures the Contractor develops a detailed programme in time, so that the baseline programme can be established earlier, and therefore the changes can be controlled and managed more efficiently.

3.3.1.2 Stakeholder Engagement

In order to produce an effective baseline programme, it is important to obtain stakeholders' engagement at the beginning of the project planning. Best practice is to hold an interactive planning workshop to engage key stakeholders as well as project managers and technical leaders. Depending on the complexity of the project, such planning workshops can last for a day to a few days. For complex projects, it is better to start with the high level programme first and then work on the detailed programme stage by stage. Key interfaces can be discussed and critical logic links can be agreed with the key parties of the project. Meanwhile, assumptions and potential schedule risks are captured during this interactive planning workshop.

3.3.1.3 Work Breakdown Structure (WBS)

Work breakdown structure is another important component for project planning because it often reflects the project scope. Establishing good work breakdown structure can effectively facilitate scope management and change management. For large international construction projects, it is worth establishing a work breakdown structure before getting into detailed activities in the planning workshop.

3.3.1.4 Schedule Quality

Due to the urgent requirement of the baseline from the Client after the awarding of the project, the project management team often focus on the delivery of the first programme but ignore the importance of the schedule quality, which may have a major impact on the project delivery later on.

There are two well-known standards used for evaluating the quality of a schedule. First, the Defense Contract Management Agency (DCMA) of the United States recommends a 14-point assessment, which includes (DCMA, 2012):

1) logic
2) leads
3) lags
4) relationship types
5) hard constraint
6) high float
7) negative float
8) high duration
9) invalid dates
10) resources
11) missed tasks
12) critical path test
13) Critical Path Length Index (CPLI)
14) Baseline Execution Index (BEI)

Deltek Acumen Fuse integrates this 14-point assessment into their schedule quality check options. The Oracle Primavera P6 also provides a schedule check option in the 8.2 version or above.

In addition, most popular quantitative schedule risk analysis software also provides a schedule quality check before undertaking any further schedule risk analysis, e.g., Primavera Risk Analysis, Deltek Acumen Risk, and Safran Risk.

3.3.1.5 Critical Path

Critical path is important for managing and controlling large international construction projects. It gives the project manager a dynamic view of when the project will be complete. It also plays a critical role in the assessment of the extension of time during the delay analysis.

The method used to assess the delay can vary in different contract forms. JCT assesses delay to the completion date required in the contract. Similar to JCT, FIDIC also assesses the delay to the relevant key dates or the completion date. As NEC specifically provides that the terminal float between the planned Completion and the Completion is owned by the Contractor, Clause 63.5 of the NEC contract provides that:

> *a delay to the completion date is assessed as the length of time that, due to the compensation event, planned completion is later than planned completion as shown on the Accepted Programme.*

3.3.2 Programme under NEC Contract

The level of detail for the programme requirements in standard forms of contract in decreasing order is NEC, FIDIC, and JCT.

NEC (New Engineering Contract) sets out comprehensive requirement and detailed procedures for the programme management, which lead to the success of large construction projects. Therefore, most of the large infrastructure projects in the UK use the NEC contract. As an example of best practice in programme management, the following section will introduce the mechanism of programme management under the NEC contract and discuss elements that lead to good front-end planning for large construction projects.

Clause 31.2 of the NEC contract sets out detailed requirements for the *Contractor*'s programme, which should include:

- *the starting date, access dates, Key Dates and Completion Date,*
- *planned Completion,*
- *the order and timing of the operations which the Contractor plans to do in order to Provide the Works,*
- *the order and timing of the work of the Client and Others as last agreed with them by the Contractor or, if not so agreed, as stated in the Scope,*
- *the dates when the Contractor plans to meet each Condition stated for the Key Dates and to complete other work needed to allow the Client and Others to do their work,*
- *provisions for float,*
  time risk allowances,
  health and safety requirements and
  the procedures set out in the contract,
- *the dates when, in order to Provide the Works in accordance with the programme, the Contractor will need*
  access to a part of the Site if later than its access date,
  acceptances,
  Plant and Materials and other things to be provided by the Client and information from Others
- *for each operation, a statement of how the Contractor plans to do the work identifying the principal Equipment and other resources which will be used and*
- *other information which the Scope requires the Contractor to show on a programme submitted for acceptance.*

The *Project Manager* then has two weeks to decide whether to accept the *Contractor*'s programme. Clause 31.3 provides four reasons why the *Project Manager* should not accept the *Contractor*'s programme:

- *the Contractor's plans which it shows are not practicable,*
- *it does not show the information which the contract requires,*
- *it does not represent the Contractor's plans realistically or*
- *it does not comply with the Scope.*

Figure 3.1 demonstrates the procedure of programme acceptance for the first programme under the NEC4 contract. For projects in which Option A or Option C applies, where an Activity Schedule is required, the *Contractor* also needs to provide information which shows how each activity on the Activity Schedule relates to the operations on the programme submitted for acceptance under Clause 31.4. Although there is no express requirement that the Activity Schedule should be identical to the activity in the programme, best practice recommends that relating Activity Schedule and activity in the programme from the beginning can improve the efficiency of the overall contract management and project control for projects under Option A and Option C.

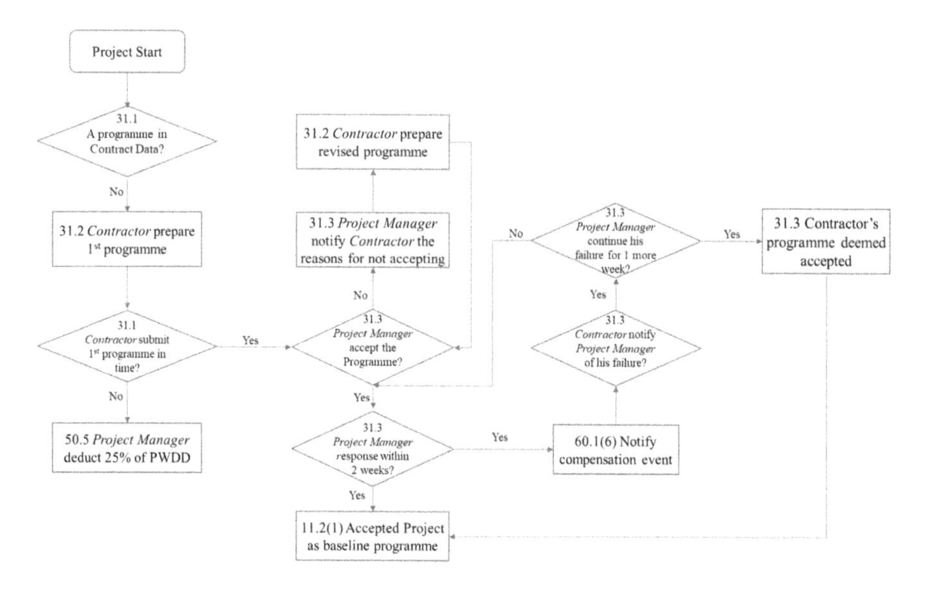

FIGURE 3.1 NEC4 procedure for acceptance of the first programme.

In order to incorporate a common issue of lack of an Accepted Programme during NEC3 implementation, NEC4 introduces the deemed acceptance of the programme to incentivize the *Project Manager* to respond to the *Contractor*'s programme on time and ensure an Accepted Programme is in place for the assessment of the compensation event.

3.3.2.1 First Programme

If the *Contractor* did not submit the first programme as required above after contract start, under Clause 50.5 of the NEC4 (2017) or Clause 50.3 of the NEC3 (2013) contract, the *Project Manager* will deduct 25% of the Price for Work Done to Date (PWDD) until the *Contractor* submits the first programme in accordance with the requirements set out in Clause 31.2.

The *Project Manager* cannot retain 25% of the PWDD for non-acceptance of the *Contractor*'s programme. Therefore, if the *Contractor* submits the programme as required by Clause 31.2, but the programme is not practicable, not realistic, or does not comply with the Scope, the *Project Manager* cannot apply the sanction under Clause 50.5.

The programme defined in the Contract Data part one during the tender stage will also waive the *Project Manager*'s right to apply sanction under Clause 50.5.

3.3.2.2 Accepted Programme

If the *Project Manager* accepted the programme, then the Accepted Programme becomes the baseline programme, which is used for the assessment of the compensation event and preparation of the project forecast as required in Clause C20.4.

Clause 11.2(1) also provides that the latest programme accepted by the *Project Manager* supersedes previous Accepted Programmes. The Accepted Programme is an on-going document updated to reflect current progress and revisions to the remaining plan. Therefore, the latest Accepted Programme acts as the latest baseline programme in the NEC contract. Each and every Accepted Programme is a very important document and goes far beyond a simple management tool, as it is used to assess change, monitor progress, and assist with early warning and compensation events.

In contrast to traditional planning practice, it keeps the baseline programme separate and updates the baseline programme when each change event is implemented. Therefore, the uncertain impact of the change event to the original baseline programme can be assessed in the decision of extension of time. Under the NEC contract, the *Contractor* needs to submit the progress programme at every interval period and submit it to the *Project Manager* for acceptance. Once the programme is accepted by the *Project Manager*, it becomes the baseline programme, which includes the actual progress to date. It is common that as the variance appears in the programme period by period, some activities may start/finish early, others may start/finish late. This may lead to a change in the critical path of the overall programme. However, as the free float to the planned Completion is shared between the *Contractor* and the *Client* on a first come, first served basis, and the extension of time is only assessed based on the planned Completion date, it may not have a negative impact on the *Client*. However, if the project under the JCT contract borrows the concept of the rolling Accepted Programme without considering other features of NEC time management, the *Client's* benefit will be undesirably impacted.

3.3.2.3 Revising the Programme

Clause 32.1 requires the *Contractor* to show in each revised programme:

- the actual progress achieved on each operation and its effect upon the timing of the remaining work,
- how the Contractor plans to deal with any delays and to correct notified Defects and
- any other changes which the Contractor proposes to make to the Accepted Programme.

Clause 32.2 requires the *Contractor* to submit a revised programme to the *Project Manager* for acceptance within the *period for reply* after the *Project Manager*'s instruction, as the *Contractor*'s intention, or within the *interval* stated in the Contract Data, which is less. The *Project Manager* then needs to decide whether to accept the *Contractor*'s programme within two weeks under Clause 31.3. The *Project Manager* cannot simply ignore each submitted programme but must either accept or give reasons for non-acceptance within two weeks. If the *Project Manager* fails to response within the required time limit, the *Contractor* can notify the *Project Manager*'s failure under Clause 31.3 of the NEC4 contract. If the *Project Manager* fails to respond for another one week, the *Contractor*'s programme is deemed to be accepted. The time associated with this deemed acceptance starts when the *Contractor* notifies the *Project Manager* of their failure to respond within the time limit; if the *Contractor* fails to notify the *Project Manager* of their failure, the deemed acceptance procedure would not start automatically. This requires both the *Project Manager* and the *Contractor* to perform in accordance with the procedure and the time limit required by the contract and sanction the non-performance Party.

If the *Contractor* disagrees with the *Project Manager*'s reason for non-acceptance, the *Contractor* should continue to revise their Accepted Programme as per the contract.

3.3.2.4 Activity Schedule

Under Option A and Option C, the contract requires the *Contractor* to submit an Activity Schedule before the contract commences. Clause A11.2(21) and C11.2(21) defines the Activity Schedule as follows:

> *The Activity Schedule is the Activity Schedule unless later changed in accordance with these conditions of contract.*

The Activity Schedule plays a significant role in the Option A contract, because the *Contractor*'s payment depends on the value of 100% completed activities to the assessment date in accordance with Clause A11.2(27). The level of detail of the Activity Schedule can have a significant impact on the *Contractor*'s cash flow; small sub-contractors may bear a significant liquidation risk if they do not set out an appropriate Activity Schedule from the beginning.

In order to claim payment for works done on time, the *Contractor* should develop its programme in sufficient detail in accordance with the requirements under Clause 31.2 and it is better to break down the activities to no longer than four weeks or the payment assessment interval. In addition, the total value in the Activity Schedule should be equal to the lump sum contract Price. Therefore, the allocation of cost in the Activity Schedule should reflect the resource and cash flow profile of the *Contractor*. Although there are no requirements for the activities in the Activity Schedule, it is good practice to relate activities in the *Contractor*'s programme to the Activity Schedule. It is worth using an advanced planning tool, such as the Oracle Primavera P6, to prepare the preliminary plan and allocate resource and cost to each activity to verify the resource and cost profile by utilizing activity priority and resource leveling to conclude an effective Activity Schedule.

Figure 3.2 shows an example of a review of the resource profile and cost profile of a project in the Primavera P6. The *Contractor* can evaluate the cashflow profile against each individual activity in order to re-sequence the programme or break down the activities further to demonstrate the realistic resource profile and cash flow profile.

Because the Option A contract is often used for the sub-contract or the professional services contract, the sub-contractor and/or professional consultant often use the resource and cost loaded P6 programme to provide associated information related to the Activity Schedule. In such circumstances, because payments largely depend on the activities in the programme, the project programme becomes one of the most important components for all levels of project management. Consequently, the programme under Option A are often well developed and progress updates are more efficient and accurate. All associated detailed records are also well kept. As a result, a good level of project control can be achieved under the Option A contract.

3.3.2.5 Sectional Completion

Large construction projects are often divided into different phases or sections. The secondary Option X5 provides the Parties to set up the sectional completion dates prior to the overall project Completion Date. Each of the section completion dates will be related to a Key Date in the programme and the impact on each change event will be evaluated not only against the Completion Date but also these Key Dates.

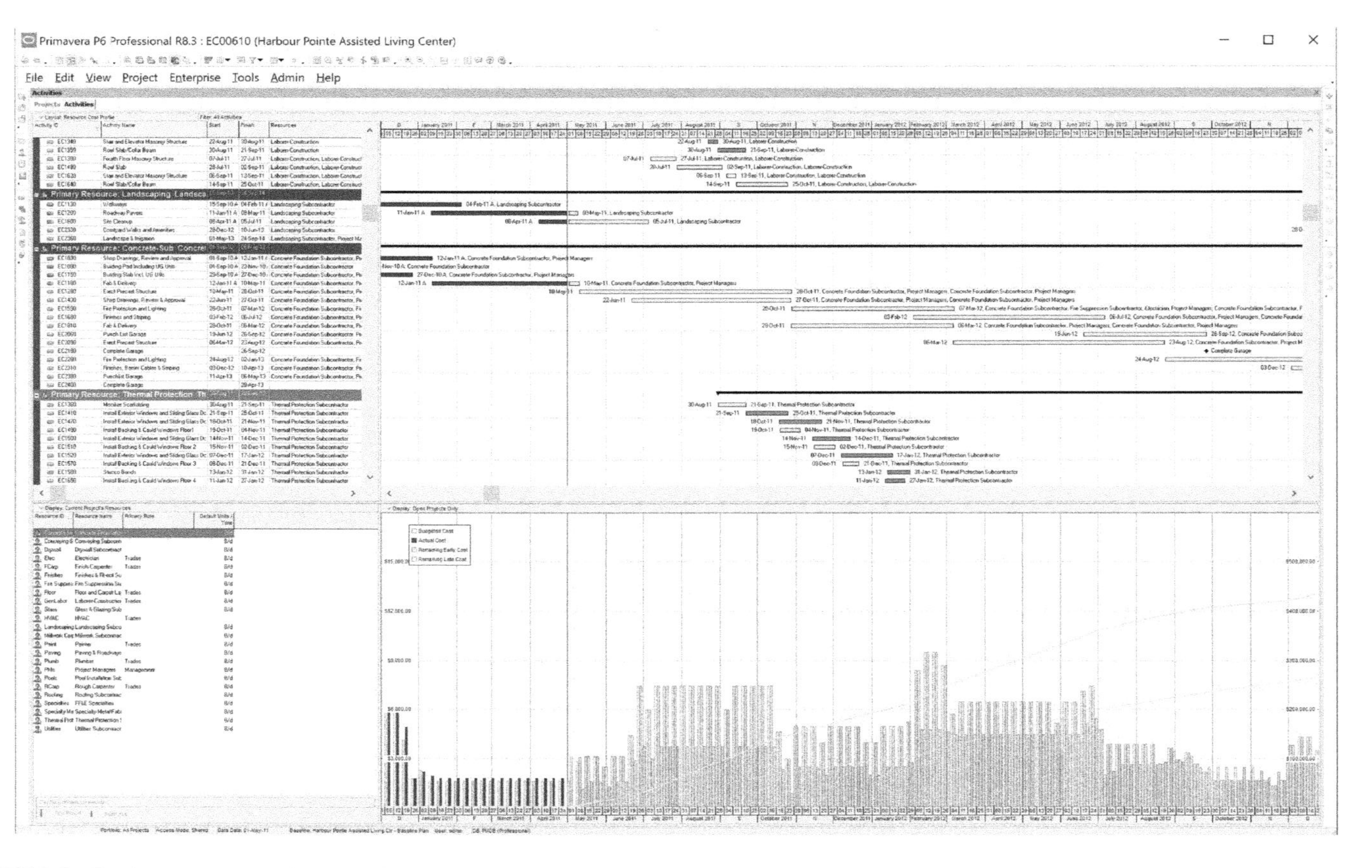

FIGURE 3.2 Project resource and cost profile evaluation under Primavera P6. (Copyright Oracle and its affiliates. Used with permission.)

3.3.2.6 Bonus for Early Completion

Under Clause 35.1, the *Client* needs to take over within two weeks after Completion unless the Contract Data states that the *Client* is not willing to take over before Completion. Once the *Client* takes over the *works*, the risk shifts to the *Client.* However, if the *Client* can benefit from early Completion, it usually includes Option X6 to pay an early Completion bonus, in order to incentivize the *Contractor*'s performance and increase the possibility of Completion on time.

Clause X6.1 provides:

> *The Contractor is paid a bonus calculated at the rate stated in the Contract Data for each day from the earlier of*
>
> - *Completion and*
> - *the date on which the Client takes over the works*
>
> *until the Completion Date.*

Similar to the delay damages, the bonus for early completion is usually defined as a daily rate in the Contract Data. In practice, the delay appears more frequently than the early completion. Therefore, this clause benefits the *Client* more. In particular, for projects that have a strict Completion Date to achieve, or if the delay would be detrimental to the *Client*'s reputation or result in significant financial loss, it would be better to incentivize the *Contractor*'s performance on the schedule through Option X6.

3.3.2.7 X7 Delay Damages

In contrast to the Option X6 bonus for early completion, Option X7 sets out the delay damages to the *Contractor* if he fails to complete by the Completion Date. Clause X7.1 provides:

> *The Contractor pays delay damages at the rate stated in the Contract Data from the Completion Date for each day until the earlier of*
>
> - *Completion and*
> - *the date on which the Client takes over the works.*

For large construction projects, the Contract Data usually defines the delay damages as a daily rate. Delay damages are often set out for each sectional completion key date in large construction projects. For smaller projects, or the contract between the main *Contractor* and the sub-contractors, the delay damages may also be defined as a weekly rate. The delay damages act as the liquidated damages, which will be discussed further in section 3.5.4.

However, if the *Client* takes over part of the works before Completion, under Clause X7.3, the delay damages will be reduced proportionally.

3.3.3 Programme under FIDIC Contract

FIDIC 1999 edition does not have comprehensive requirements for the project programme like NEC, but the 2017 edition sets out a detailed procedure and requirements for the project programme, in order to improve time management and control of the overall project.

3.3.3.1 General Programme Requirement

Sub-Clause 8.3 "Programme" provides the detailed requirements for the Contractor's programme as follows:

- Commencement date, Time for Completion and sectional completion date
- the access date and date of possession the Site
- All activity showing early start/finish date, late start/finish date, float and critical path
- Key delivery date of Plant and Materials
- the sequence to carry out the works including anticipated duration of each stage for:
  - design,
  - preparation and submission of Contractor's Documents,
  - procurement,
  - manufacture,
  - inspection,
  - delivery to Site,
  - construction,
  - erection,
  - installation,
  - work to be undertaken by any nominated sub-contractor
  - testing,
  - commissioning, and
  - trial operation.
- Review period
- Sequence and duration for inspection and test
- Supporting report includes description of major stages of the work, the construction method, estimate to the Contractor's Personnel and Equipment.

In addition, under FIDIC 2017, the Contractor is required to use the mandatory programming software for project planning which is specified by the Employer. In the absence of the Employer's requirement, the Engineer will decide the acceptable software. Nevertheless, most large international construction projects undertaken under the FIDIC contract use Primavera P6 as the planning tool to develop the programme.

3.3.3.2 Revised Programme

For the revised programme, the following elements are also required:

- remedial works of identified defects
- actual progress to date and any delay together with associated knock-on effect
- supporting report including any significant changes to the previous programme and the Contractor's proposal to mitigate the delay

Although resource allocation of the programme is not expressly required in the contract, in order to achieve the requirement for detailed estimation of the Contractor's

Personnel and Equipment for each major execution stage under Sub-Clause 8.3.(k).(iii), the Contractor ought to undertake resource loading in the programme for an effective planning and reporting purpose.

3.3.3.3 Process and Time Limit for Programme Submission and Acceptance

The Contractor needs to submit the first programme within 28 days of receiving the Notice for Commencement of Works under Sub-Clause 8.1. Then the Engineer needs to give the Contractor Notice within 21 days if the programme does not comply with the Contract.

FIDIC does not provide a specific time interval for the Contractor to submit the subsequent revised programme, only provides "*whenever any programme ceases to reflect actual progress or is otherwise inconsistent with the Contractor's obligations.*" However, under Sub-Clause 4.20.(h), the monthly progress report requires the Contractor to compare the actual progress to planned progress and provide details of areas of adverse delay together with the method to overcome the delay. Furthermore, in the Contractor's Application for Interim Payment under Sub-Clause 14.3(c), the Contractor is required to submit a Statement to the Engineer including the detailed amount that the Contractor is entitled to with supporting details of progress in accordance with the Progress Report under Sub-Clause 4.20. Therefore, although there is no a specific progress update interval, to comply with the monthly progress report and interim application for payment requirements, the Contractor ought to update its programme at least once a month. The Engineer can also give Notice to the Contractor for a revised programme at any time, if they believe the Contractor's preprogramme is inconsistent with actual progress or the Contractor's obligations.

Upon receiving the Contractor's programme, the Engineer needs to response within 14 days if the programme does not comply with the contract or stop to reflect the actual progress. Figure 3.3 explains the process of the programme under FIDIC 2017 edition and associated time limit for the Contractor and the Engineer.

If the Engineer fails to serve a Notice of No-objection within 21 days of submission of the first programme or 14 days of the revised programme, then the Contractor's programme will be deemed accepted as the Programme. Then the Contractor needs to proceed in accordance with the Programme.

Sub-Clause 1.1.66 of the FIDIC 2017 Red Book, Sub-Clause 1.1.67 of the FIDIC 2017 Yellow Book, and Sub-Clause 1.1.57 of the FIDIC Silver Book provide the definition of "Programme" as:

> *a detailed time programme prepared and submitted by the Contractor to which the Engineer has given (or is deemed to have given) a Notice of No-objection under Sub-Clause 8.3 [Programme].*

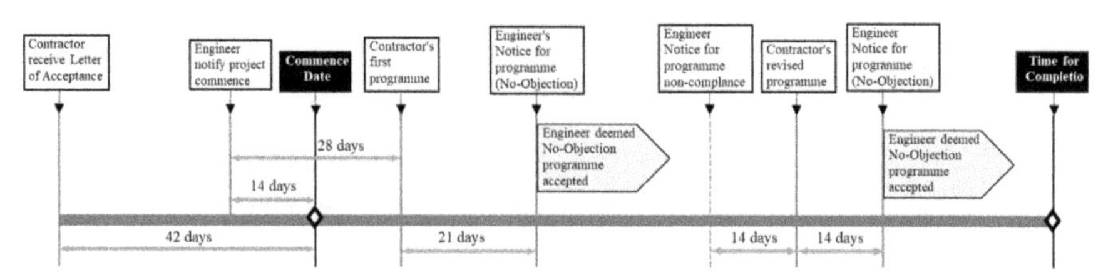

FIGURE 3.3 Process and Time Limit Programme under FIDIC 2017 forms of Contract.

Consequently, FIDIC 2017 forms of contract require substantial planning engagement, and the Contractor needs to develop and maintain a detailed programme in order to achieve all the requirements under the contract. Although this will result in increased cost for the project planning and project control personnel, the process leads to a comprehensive programme set up from the project start and the reporting of potential issues in time. This ensures effective project control undertaken from the front-end of the project and would fundamentally improve the project performance as NEC does.

3.3.3.4 Delay Damages

Like other standard forms of contract, FIDIC also provides delay damages under Sub-Clause 8.8. The Delay Damages under FIDIC forms of contract will be calculated at a daily rate, for the period between the Time for Completion and the Date of Completion for the whole Works or by Section. The total of Delay Damages will be capped by the maximum value defined in Section 8.8 of the Contract Data.

In order to couple with the potential challenges of the liquidated damages constituting a penalty under common law jurisdiction, both the daily rate of Delay Damages and the maximum value under the FIDIC contract are sometimes also defined as a percentage of the total contract price.

3.3.4 Time Management under JCT Contract

In contrast with FIDIC and NEC, the time management under the JCT contract usually does not rely on a comprehensive planning programme, but rather depends on the procedure set out in the contract to mainly manage the sectional completion date or Completion Date. In *Greater London Council v. The Cleveland Bridge and Engineering Co Ltd and Another* (1986) 34 BLR 50, although the Contractor left part of the work to finish just before the Completion Date, and the Employer was then liable to pay a large value of fluctuation which could be avoided if the Contractor arranged its work appropriately, as per the Employer's argument. The court held that the Contractor can perform in any way it prefers as long as it completes the works on time, and the Contractor's action was not a breach of contract. Therefore, under the JCT contract, the Contractor can organize the works in any sequence as it wishes, as long as it finishes the works by the Completion Date.

3.3.4.1 Programme

The JCT contract does not specifically require a programme or schedule as part of the contractual document. The contract simply sets out the "Date for Possession of the Site" in Section 2.3, and the "Date for Completion of the Works" in Section 1.1 of the contract particulars or the pre-agreed adjustment under Clause 2.25. JCT DB 2016 does not set out any specific requirement for a Contractor's programme.

Although Clause 2.9.1.2 of JCT SBC 2016 requires the Contractor to provide a "*master programme for the execution of the Works identifying the critical paths*," and the Employer can set out additional requirements for the Contractor to provide a programme under other forms of JCT contract, the programme is not included in the contract documents, and programme is referred in any other terms, hence the Contractor has no obligation to the programme.

3.3.4.2 Progress

Clause 2.3 of JCT DB 2016 requires the Contractor to conduct the works "regularly and diligently," in order to achieve the Completion Date. In *West Faulkner Associates v. Newham London Borough Council* CA 16 Nov 1994, Lord Justice Simon Brown in the Court of Appeal defined the words "regularly and diligently" as follows:

> *Taken together the obligation upon the Contractor is essentially to proceed continuously, industriously and efficiently with appropriate physical resources so as to progress the works steadily towards completion substantially in accordance with the contractual requirements as to time, sequence and quality of work.*

Clause 2.25 further requires the Contractor to "*constantly use his best endeavors to prevent delay in the progress of the Works*" and "*do all that may reasonably be required to the satisfaction of the Employer to proceed with the Works.*"

Although there is no specific programme that requires the Contractor to complete certain works period by period, and JCT does not specify when a programme requires revision nor set out any submission timelines, the Contractor is obliged to undertake self-control to achieve the contract Completion Date.

3.3.4.3 Relevant Event

Similar to the compensation event of the NEC contract, JCT DB 2016 sets out 14 Relevant Events to allow the Contractor to claim extension of time under Clause 2.26:

1) Change to the scope of work
2) Employer's instruction
3) Delay giving the Contractor of site possession
4) Archaeological objects
5) Suspension by the Contractor, e.g., no payment
6) Any prevention or hinderance by the Employer or its representative
7) External body carrying out works to proceed its statutory obligation
8) Adverse weather conditions
9) Loss or damage arise by Specified Perils, e.g., insurance
10) Civil commotion or the use or threat of terrorism
11) Strike, lock-out as such
12) New legislation
13) Delay by the authorities for necessary permission or approval
14) Force majeure

Please note that the Contractor cannot claim extension of time due to the poor performance of an Employer-nominated sub-contractor.

In contrast with the NEC early warning notification, under the JCT contract, the Contractor can only claim additional time and cost after the risk event has occurred. Due to the lack of programme, the extension of time will be assessed on the Completion Date. Therefore, if a Relevant Event delays the Completion Date for a certain duration, then the Contractor is entitled to extension of time for the relevant duration. However, due to the lack of programme, it is unlikely to assess the critical path. The actual extension of time is either grounded based on the time required for each individual Relevant Event or assessed backward based on the actual duration taken in the construction implementation.

3.3.4.4 Acceleration

Like NEC, there are two possible situations to change the Completion Date set out in the initial contract: through the extension of time to push the Completion Date backward or through the acceleration to bring the Completion Date forward.

The Employer can request the Contractor to submit a quotation for acceleration under Section 4 of the Schedule 2 supplemental provisions. The Contractor then must provide a proposal either with a quotation within 21 days with his estimate of the time that can be saved, and additional cost required, or a reason that the acceleration is impractical in accordance with Section 4.1.1 of the Schedule 2 supplemental provisions. The Employer then needs to either accept or reject the Contractor's proposal within 7 days. If the Employer accepts the Contractor's proposal, then the Completion Date will be brought forward under Section 4.3.1, and the contract sum will be increased accordingly under Section 4.3.2. However, if the Employer rejects the Contractor's proposal, then the Contractor is entitled to payment for preparing the proposal under Section 4.4.1. Figure 3.4 demonstrates the process of acceleration.

3.3.4.5 Procedure

If it becomes reasonably apparent that the progress of the Works is to be delayed, under Clause 2.24.1, the Contractor must notify the Employer of the causes of the delay and identify a Relevant Event under Clause 2.26. The Contractor is then required to provide an estimation of the expected delay as soon as possible under Clause 2.24.2 and notify any changes of such a delay estimation thereafter at a reasonable time.

In contrast to NEC and FIDIC, the Contractor's notification of a Relevant Event is not the condition precedent to its claim in accordance with the judgment made in *London Borough of Merton v Stanley Hugh Leach Ltd* (1985) 32 BLR 51 (Ch). In addition, JCT does not set a specific time bar for the Contractor to claim a Relevant Event. Therefore, many claims under the JCT contract are accumulated to the back-end of the project or proceed via dispute resolution post project completion.

Upon receiving the Contractor's notice, the Employer should grant extension of time within 12 weeks, if it is a valid Relevant Event and it is likely to be the cause of

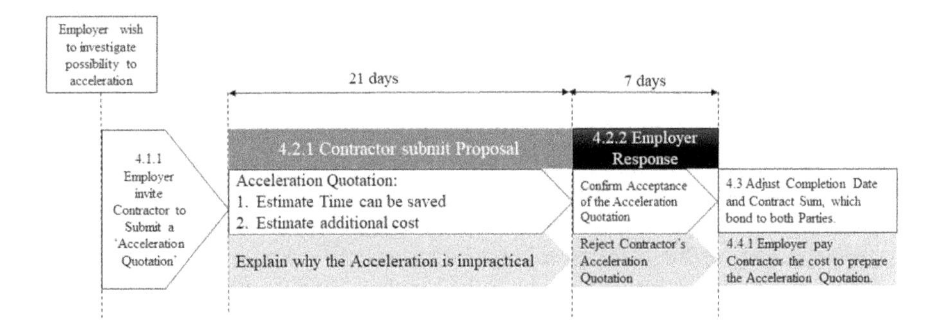

FIGURE 3.4 Acceleration procedure under the JCT contract.

the delay of Completion Date under Clause 2.25. Then the new Completion Date is fixed in accordance with Clause 2.25.4.

3.3.4.6 Liquidated Damages

Under Clause 2.3, the Contractor is obliged to complete the works by Completion Date. If the fixed Completion Date is provided in the contract and adjusted in accordance with extension of time under Clause 2.26 or acceleration under paragraph 4.1 of Schedule 2, the Employer retains their right to claim liquidated damages when the Contractor fails to complete the works by the fixed Completion Date. The rate of liquidated damages is usually defined on a weekly or daily basis in Section 2.29.2 of the contract particulars. The liquidated damages cannot only be set up for the Completion Date of the entire works, but also for the sectional Completion Date.

3.4 DELAY

If the Contractor fails to complete the works by the completion date, then the duration between the contractual completion date and the actual completion date is used to calculate the delay damages, which often become the critical area of dispute between the parties. Due to the complex nature of large international construction projects, over two-thirds of construction projects finish after the contractual completion date. Over the years, although the technology has developed significantly, the percentage of time overrun hasn't improved much. Many facts can contribute to the delay of a construction project. The most common causes of delay include: change of scope and/or design, lack of key resources, ineffective management, force majeure or adverse weather conditions, unforeseen ground conditions, and lack of collaboration.

The legal principles of delay are described in Figure 3.5. As discussed in Section 3.2, the Contractor may oblige to complete the works under express and/or implied terms. If the Employer prevents the Contractor from performing during the project implementation, the Contractor is then entitled to extension of time. Failure to grant the extension of time by the Employer would set the time at large and the Contractor can complete the work in a reasonable time. If proper extension of time is granted, the completion date is reset and the liquidated damages remedy is preserved. If the Contractor fails to complete the works by the completion date, the Employer can claim damages either through delay damages if it is set out in the contract, or the general damages in the absence of the liquidated damages provisions. The following sections will explain each individual concept in detail.

3.4.1 Types of Delay

If the Contractor fails to complete the project by the contract completion date, then the variance between actual completion date and the contractual completion date is the amount of delay occurred. The Employer is entitled to claim damages associated to the delay. Due to the scale and value of large construction projects, the delay damages can be of significant value to both the Employer and the Contractor. Hence, it often becomes the key area of dispute in large construction projects.

When an analysis of the amount of delay in a project is undertaken, the critical path is crucial, because it drives the completion date and only the delays affected by

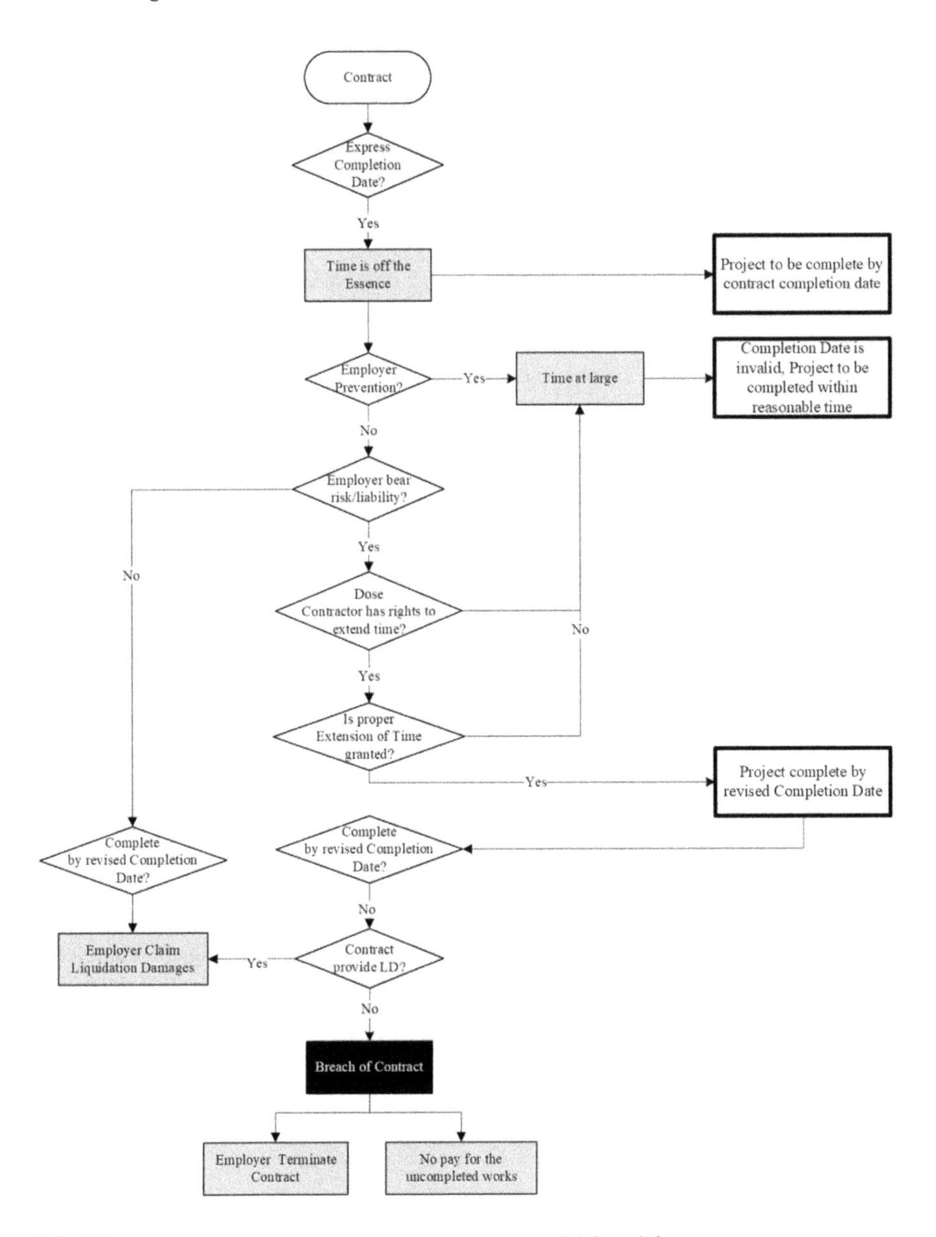

FIGURE 3.5 Legal principle for time management and delay claim.

the completion date will give rise to a claim of damages. Consequently, delay can be classified as critical delay and non-critical delay. Non-critical delay appears on the activities with sufficient float to resolve the potential impact to the completion date. Critical delay is a delay that changes the completion date. It is not necessarily a delay that occurred on the existing critical path, but the consequence of the delay will result in changing of the critical path. In large construction projects, there are many network paths linking the activities from start to end; the critical path usually determines the completion date of the project. The non-critical paths with low float may become the critical path if some delay arises which exceeds the existing float. Therefore, the project control team should keep detailed records for each programme and undertake impact analysis for each change event, because the accumulative impact can result in the change of float as well as the critical path in the future event.

The critical delay may lead to a significant commercial impact, e.g., liquidated damages. Therefore, the question of who is liable for a delay often becomes a key concern for the contract parties.

The critical delay is usually classified as "Excusable delays" and "Non-excusable delays," in accordance with the party responsible for the delay event. The Non-excusable delay is at the Contractor's risk, and the remaining are excusable delays. Based on the Contractor's right to claim the delay event, delay can also be categorized as compensable delay and non-compensable delay. All non-excusable delay is non-compensable, as such delay relates to delay events for which the Contractor bears the risk and liability. However, excusable delay can be compensable if the Employer bears the risk and liability; or non-compensable if the delay is caused by a natural event, such as adverse weather conditions or a third-party action.

Table 3.2 explains the types of delay and the Contractor's entitlement under each scenario. For the compensable delay, the Contractor is entitled to an extension of time and claim for damages associated to the delay event. For the non-excusable delay, the Employer is entitled to claim liquidated damages for the period of delay, if the contract provides liquidated damages entitlement, or general damages otherwise. For the natural risk event, which constitutes the non-compensable but excusable delay, the risks are shared between the Contractor and the Employer. The Employer takes the risk of time and the Contractor takes the risk of cost; although the Contractor is entitled to the extension of time, but the cost and profit are not recoverable.

3.4.2 Extension of Time (EOT)

Most standard forms of contract contain express terms for adjusting the completion date under defined circumstances, for instance Clause 2.26 under JCT DB 2016, Sub-Clause 8.5 under FIDIC 2017, and Clause 60.1 under NEC4. On the one hand, an extension of time clause allows the Contractor to claim extra time to complete the works where appropriate and provides relief from paying liquidated damages or

TABLE 3.2
Types of Delay

Type of Delay	Excusable		Non- Excusable
Causation	Employer Risk Event	Natural Event	Contractor Risk Event
Compensation	Compensable	Non-Compensable	Non-Compensable
Responsible	Employer	Employer & Contractor	Contractor
Damages for Contractor	EOT and/or Cost	EOT & no Cost	No
Damages for Employer	No	No	Liquidated Damages or loss of operation/profit

general damages. On the other hand, an extension of time clause allows the Employer to grant additional time for the Contractor to complete the works when the Employer is responsible for the event causing the delay. This ensures the contract completion date is reset to a later fixed date and the Employer retains its right to claim liquidated damages if the Contractor fails to complete on the fixed contractual completion date. Without the EOT provision, when the works is delayed by the Employer's risk event, the fixed completion date will be set at large. For example, in *Wells v Army & Navy Cooperative Society* (1902) 86 LT 764, the Employer lost its entitlement to claim liquidated damages because the Employer failed to give the Contractor access to the site.

As demonstrated in Table 3.2, the Contractor is entitled to claim extension of time under two circumstances: first, if the delay is caused by an event which is the responsibility of the Employer; second, the delay is caused by a natural event for which the Contractor is not liable under the contract, for example force majeure, strike, or adverse weather conditions. When the contract is silent on the event that the Employer is liable for, the delay of such event will be at the Contractor's risk.

The standard forms of contract refer to different terminology for the Contractor's entitlement of compensation. The NEC contract refers to "compensation event" and Clause 60.1 lists out 21 elements that the *Contractor* is entitled to for compensation as well as extension of time, plus the *Client*'s liability events listed in Clause 80.1 and additional *Employer*'s risks included in Contract Data part one. Under the NEC contract, all compensation events are entitled to both time and cost compensation. This will be discussed further in Chapter 6, Change Management.

The JCT Standard Building Contract (SBC) 2016 refers to Relevant Events with 15 sub-categories set out under Clause 2.29.

The prevention events are set out in Clause 2.29.7 as follows:

> *any impediment, prevention or default, whether by act or omission, by the Employer, the Architect/Contract Administrator, the Quantity Surveyor or any Employer's Person, except to the extent caused or contributed to by any default, whether by act or omission, of the Contractor or any Contractor's Person.*

FIDIC does not provide a specific terminology but defines the Contractor's rights for extension of time in various clauses in the contract. The entitlement of extension of time claim largely depends on the risk allocation strategy under the contract. Table 3.3 summarizes the common entitlements of extension of time under JCT, NEC, and FIDIC forms of contract.

In addition, the FIDIC contract specifically provides Contractor's entitlement for extension of time for setting out error (Sub-Clause 4.7), specified perils (Sub-Clause 4.7), fossils (Sub-Clause 4.24), interference with test on completion (Sub-Clause 10.3), and changes in legislation (Sub-Clause 13.7).

3.4.3 Delay Analysis

When a claim or dispute arises associated to the late completion, the delay analysis is often used to establish the damages and/or the settlement for the claim. The standard principle and methods that are used for undertaking delay analysis have

TABLE 3.3
Entitlement of the Extension of Time under Standard Forms Contract

EOT Entitlement	JCT	NEC	FIDIC
Terminology	Relevant Event	Compensation Event	
Variations	√	√	√
Instructions	√	√	√
Possession/Access Site	√	√	√
Employer's Delay	√	√	√
Authority Delay	√	√	√
Adverse Weather Conditions	√	√	√
Force Majeure	√	√	√
Suspension	√	√	√
Unforeseen Physical Conditions	√	√	√
Changes in Legislation	√	√	√
Specified Perils	√		
Fossils			√
Interference with Test on Completion			√
Setting out Error			√

been provided in the Society of Construction Law Delay and Disruption Protocol and Forensic Schedule Analysis by the American Association of Cost Engineers (AACE). Figure 3.6 classifies the different delay analysis methods based on their specific features.

According to the point of time, which determines the impact of the delay, the analysis can be classified as prospective analysis which is undertaken prior to the delay event occurring and retrospective analysis which is performed after the delay event. The retrospective analysis can then be classified into the observation

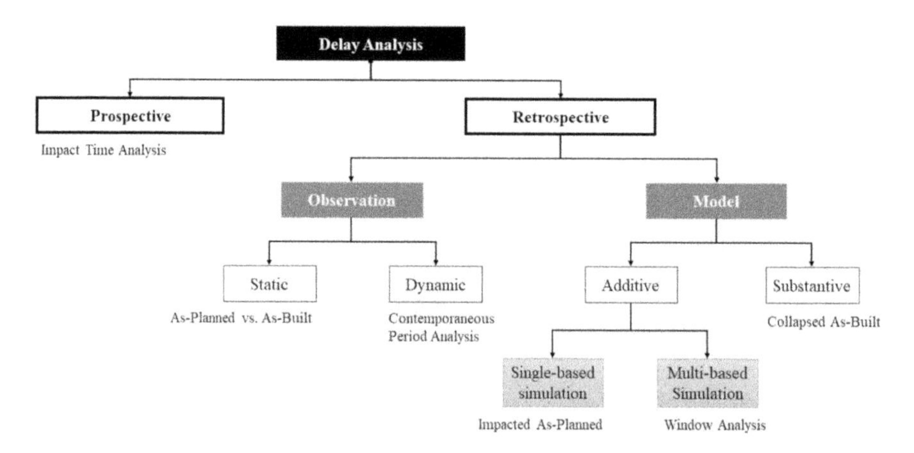

FIGURE 3.6 Delay analysis methods under AACE (2011) recommended Forensic Schedule Analysis.

method and the modeling method. The observation method relies on the available programme, including baseline programme, As-Built programme and progress programme. While the static method simply compares the As-Planned programme with the As-Built programme the dynamic method involves comparing the programme change on a period by period basis. Unlike the observation approach, which merely reviews the programme, the model approach analyzes the programme activity in detail, and may add additional activity to demonstrate the delay events or subtract certain activities to investigate different scenarios. The additive model is based on a single baseline programme to add multiple delay events to analyze the impact as per the Impacted As-Planned analysis, or simulate with different scenarios as some window analyses do.

Whereas the Society of Construction Law structured the delay analysis methods differently in its Delay and Disruption Protocol Delay (SCL, 2017) as shown in Figure 3.1, SCL (2017) summarizes the six types of delay analysis as follows:

1) Impacted As-Planned Analysis
2) Time Impact Analysis
3) Time Slice Windows Analysis
4) As-Planned vs. As-Built Window Analysis
5) Retrospective Longest Path Analysis
6) Collapsed As-Built Analysis

Like AACE, SCL first classifies the delay analysis methods based on whether the delay impact is determined prospectively or retrospectively. The Prospective analysis identifies the causes first and then derives the effect. Whereas, the Retrospective analysis evaluates the effect of the cause first and then investigates the causes. Furthermore, SCL also classifies the delay analysis based on how the critical path has been determined, whether it's prospectively, contemporaneously, or retrospectively as shown in Figure 3.7.

In combination of the AACE and SCL's practice recommendations, the following delay analysis methods will be discussed in more detail:

a) As-Planned vs. As-Built
b) Impacted As-Planned
c) Time Impact
d) Collapsed As-Built
e) Window Analysis

Each delay analysis approach has advantages and disadvantages. The selection of actual approaches used for the specific cases depends on the circumstances of each case.

3.4.3.1 As-Planned vs. As-Built

This approach is to compare the Contractor's original planned programme, typically the baseline programme against the As-Built programme, which is the schedule that actually occurred to complete the work. This analysis can be undertaken relatively easily by using comprehensive planning software, such as Oracle Primavera P6

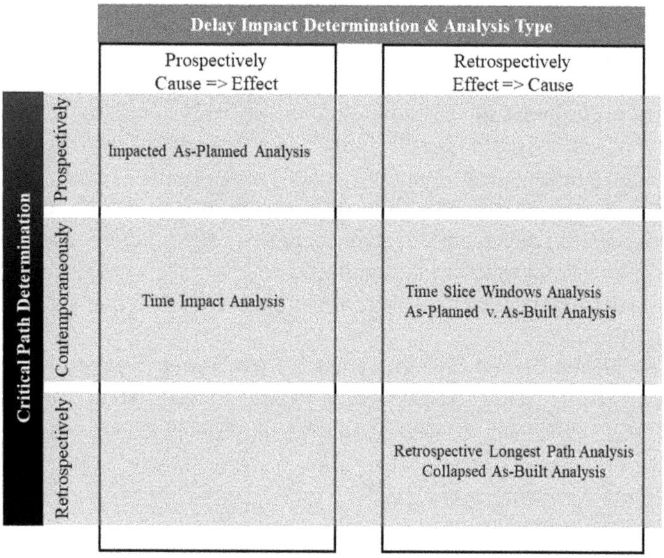

Delay Impact Determination & Analysis Type
Prospectively
Cause => Effect
Retrospectively
Effect => Cause
Critical Path Determination
Prospectively
Contemporaneously
Retrospectively
Impacted As-Planned Analysis
Time Impact Analysis
Time Slice Windows Analysis
As-Planned v. As-Built Analysis
Retrospective Longest Path Analysis
Collapsed As-Built Analysis

FIGURE 3.7 Classification of delay analysis methods under the SCL Delay and Disruption Protocol (SCL, 2017).

Enterprise Project Portfolio Management, Microsoft Project, etc. Figure 3.8 demonstrates an example of As-Planned vs. As-Built analysis.

In addition, to understand the potential practical challenges beyond the activity bar chart, the resource and cost loaded programme can further provide a comparison graph of labor, cash flow, plant, and equipment to assist in-depth analysis of the delay. The starting point of this analysis is the accepted baseline programme; if the Contractor is entitled to extension of time under the contract, the last Accepted Programme before each delay event occurs will be used for the As-Planned programme. If the original baseline programme is unrealistic, or the relevant EOT

As Planned (Baseline Programme)
As-Built (Actual Programme)
Delay
Delay Event

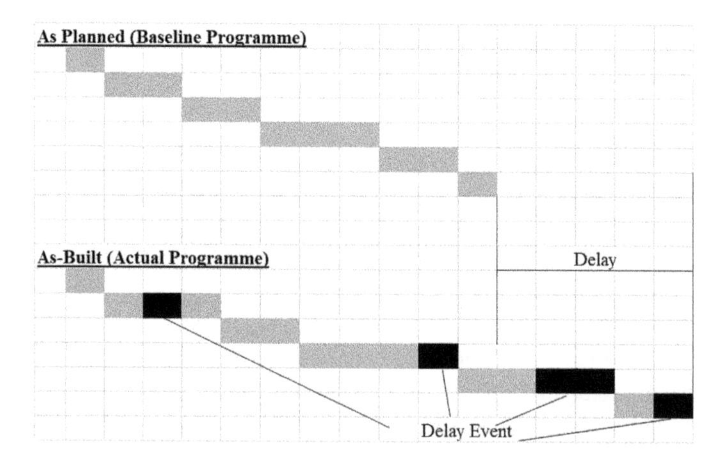

FIGURE 3.8 As-Planned vs. As-Built Analysis.

hasn't been built into the revised baseline programme appropriately, it will impact the result of the analysis.

3.4.3.2 Impacted As-Planned

Impacted As-Planned analysis measures the impact of the delay event on the Contractor's planned schedule. Under this approach, different delay events are categorized as activities and added to the existing baseline programme in chronological order. The impact of each type of delay is recorded in the programme together with the responsible parties. The impact can be identified by comparing the completion date before and after the delay event. Meanwhile, this approach is good for analyzing the concurrent delay and finalizing the contract completion date. When building in all the delay events to which the Contractor is entitled for extension of time in the As-Planned programme, the impact of the contract completion date can be determined, as shown in Figure 3.9.

Large international construction projects usually have complex programme. When multiple delay events impact the programme from different angles, the float as well as critical path may change. When determining the final contract completion date, the parties should consider the sequence in which the delay events occurred when utilizing the total float, because this float is claimed based on a first come, first served principle. Furthermore, both the *Contractor* and the *Client* should bear in mind that under the NEC contract, the *Contractor* owns the time risk allowance and terminate float.

3.4.3.3 Time Impact Analysis

The Time Impact analysis takes the Contractor's baseline programme as the starting point. The baseline programme is then updated with actual progress to the data date. The preliminary progress programme clearly records that any delay is the responsibility of the Contractor or the Employer respectively. The planner then develops the forecast programme at the point of update, including any known delay events that are the responsibility of the Employer. The total delays that are not the responsibility of the Contractor will be its entitlement for the extension of time, as shown in Figure 3.10.

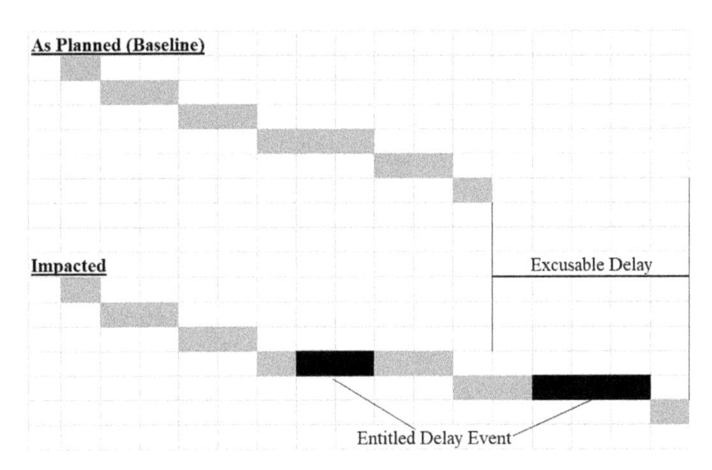

FIGURE 3.9 Impacted As-Planned Analysis.

3.4.3.4 Collapsed As-Built

Collapsed As-Built analysis is a retrospective approach. It begins with the As-Built programme. It then identifies both the Contractor's delay event and the Employer's delay event, and finally removes the delay events for which the Contractor is not responsible. The remaining programme shows when the Contractor would complete the project "But For" the Employer's delay. The difference between the completion date of the adjusted As-Built programme and the actual completion date is extension of time that the Contractor is entitled to. It is also known as "As-Built But For" analysis, as shown in Figure 3.11.

Because the actual dates are fixed when updating the As-Built schedule in the planning software, e.g., Oracle P6 or Microsoft Project, when using the computer software to undertake this analysis requires the planner to convert the As-Built programme to an un-actualized programme in order to obtain the appropriate "As-Built But For" programme after scheduling the programme.

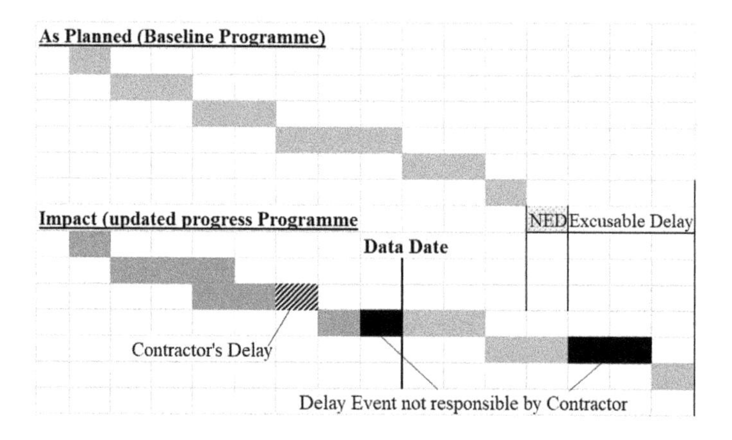

FIGURE 3.10 Time Impact Analysis.

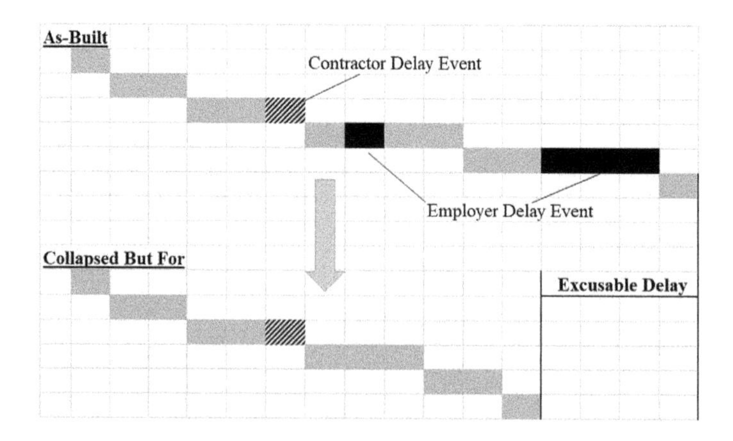

FIGURE 3.11 Collapsed But For Analysis.

3.4.3.5 Window Analysis

There are two types of window analysis, first is the time slice analysis and second is the As-Planned vs. As-Built window analysis. Both methods divide the whole project into multiple windows. It takes a snapshot of a period in which a particular delay event occurs and undertakes detailed analysis from the point taken. The time slice analysis takes the various versions of the baseline programme and updates the progress programme based on each time a change is undertaken to establish a complete programme. It analyzes the critical path in each version of the programme and establishes a more accurate programme for the project. A record is generated each time when a delay event arises, and the total of the delay events that are not the responsibility of the Contractor is then added to the Contractor's entitlement to the extension of time for the whole project.

As shown in Figure 3.12, the lattice bar on the top is the As-Planned baseline programme, the black bar is the delay event that occurred in each separate window. The grey bar at the bottom plus the total of the red bars would be the total As-Built duration, and the total of the black bars is the Contractor's entitlement to extension of time.

The window analysis can provide the most accurate results, and it considers the dynamic nature of the critical path in each window undertaken by the analysis. However, such comprehensive analysis is usually time consuming and may cost much more than other methods.

Determining which method is best to use relies on the information available and the intended budget.

Just as different techniques require difference information, the results achieved may also be different. In *Balfour Beatty Construction Ltd. v. The Mayor and Burgesses of the London Borough of Lambeth* [2002] EWHC 597 [TCC], the adjudicator's decision was challenged as the technique was not afforded commentary by the respondent. Consequently, the respondent's position was upheld. *Ascon Contractors Limited v. Alfred McAlpine Construction Isle of Man Limited* (1997 ORB-361 & 1998 ORB-315) also demonstrated that the wholly theoretical calculations will be unlikely to succeed.

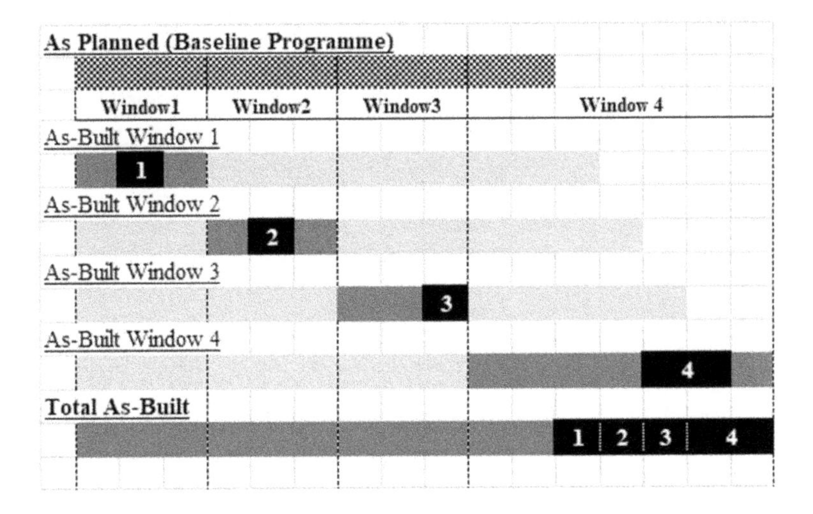

FIGURE 3.12 Window Analysis.

3.5 BURNING ISSUES

There are some issues associated with the delay dispute that often become the critical focus of the parties, in particular, the prevention principle may potentially set the time at large and de-place the liquidated damages. Furthermore, issues also arise in the event of concurrent delay, and the ownership of different type of floats when undertaking delay analysis and assessing the entitlement of the extension of time. The following section explains the legal principles for these burning issues and elaborates the current position of the relevant law.

3.5.1 Prevention Principle

The fundamental rule of the prevention principle is that the party cannot benefit from its own wrong. The Employer cannot insist that the Contractor achieve the contract completion date if they have hindered the Contractor's performance, as the Employer may benefit from the liquidated damages by their own default. This is known as the "prevention principle" in English common law. The "prevention principle" can be traced back to Comyn's Digest in 1762. In condition L(6), it provides that if the performance of a party to the contract is prevented by the other party, they are not liable for such a default. The rule of law of the prevention principle was preliminarily laid down in the Victorian case *Holme v Guppy* (1838) 150 E.R. 1195, where Parke B at paragraph 1196 noted:

> *if the party be prevented, by the refusal of the other contracting party, from completing the contract within the time limited, he is not liable in law for the default. ...The plaintiffs were therefore left at large; and consequently, they are not to forfeit anything for the delay.*

This principle has been developed over the centuries and the current application was established from the English leading case of *Peak Construction (Liverpool) Limited v McKinney Foundations* [1970] 1 BLR 111. In this case, part of the 58-week delay was caused by the Employer's breach. Salmon LJ noted in paragraph 121 that if the Employer contributed to failure to complete on time, without extension of time, 1) the Contractor is no longer bound to the original time limit of completion and time becomes at large and 2) the Employer cannot claim liquidated damages.

Further, in *Dodd v Churton* [1897] 1 Q.B. 562, Lopes LJ stated in paragraph 568 that:

> *It has been often laid down that, where there is provision that a contractor shall pay penalties for delay as in the present case, no penalty can be recovered where delay has been occasioned by the act of the person endeavouring to enforce the penalties.*

In *Trollope & Colls v North West Metropolitan Regional Hospital Board* [1973] 1 WLR 601, Lord Denning stated that:

> *(1) ...when there is a stipulation for work to be done in a limited time, if one party by his conduct...renders it impossible or impracticable for the other party to do his work within the stipulated time, then the one whose conduct caused the trouble can no longer insist upon strict adherence to the time stated. He cannot claim any penalties or liquidated damages for non-completion in that time.*

> *(2) The time becomes at large. The work must be done within a reasonable time — that is, as a rule, the stipulated time plus a reasonable extension for the delay caused by his conduct.*

In *Multiplex Constructions (UK) Ltd v Honeywell Control Systems Ltd* [2007] EWHC 447 (TCC), Mr. Justice Jackson provides that:

> *if one party by his conduct...such as ordering extra work - renders it impossible or impracticable for the other party to do his work within the stipulated time, then [he] can no longer insist upon strict adherence to the time stated. He cannot claim any penalties or liquidated damages for non-completion in that time.*

In paragraph 56, Jackson J derives three rules for the prevention principle as follows:

> *(i) Actions by the Employer which are perfectly legitimate under a construction contract may still be characterised as prevention, if those actions cause delay beyond the contractual completion date.*
> *(ii) Acts of prevention by an employer do not set time at large, if the contract provides for extension of time in respect of those events.*
> *(iii) In so far as the extension of time clause is ambiguous, it should be construed in favour of the Contractor.*

The prevention principle is not only valid in the delay event for which the Contractor is not responsible, including Employer at fault, Employer responsible, force majeure; but also applies where the Employer fails to grant the variation and no contract mechanism for contract to obtain extension of time. The most significant impact of the prevention principle is that the Employer loses his rights to claim liquidated damages due to time set as large. In *Peak v McKinney*, the court provides that "*the liquidated damages and extension of time clauses in printed forms of contract must be construed strictly contra proferentem.*" Therefore, the Employer should provide sufficient extension of time provisions covering potential Employer's preventions, in order to retain their right for liquidated damages.

Although the rule of English law established strong grounds for the prevention principle, it can be excluded by express terms. For example, in the recent case of *North Midland Building Ltd v Cyden Homes Ltd* [2018] EWCA Civ 1744, the prevention principle is derived in the implied terms but has been excluded by the express term in the contract.

3.5.2 Time at Large

Time "at large" refers to the default position of a "reasonable time" for completion as opposed to a fixed date.

The time may be set at large in the following circumstances:

1) there is no contractual completion date
2) a delay is not the responsibility of the Contractor and no extension of time is granted to the Contractor

Figure 3.13 illustrates the situations where the time will be set at large.

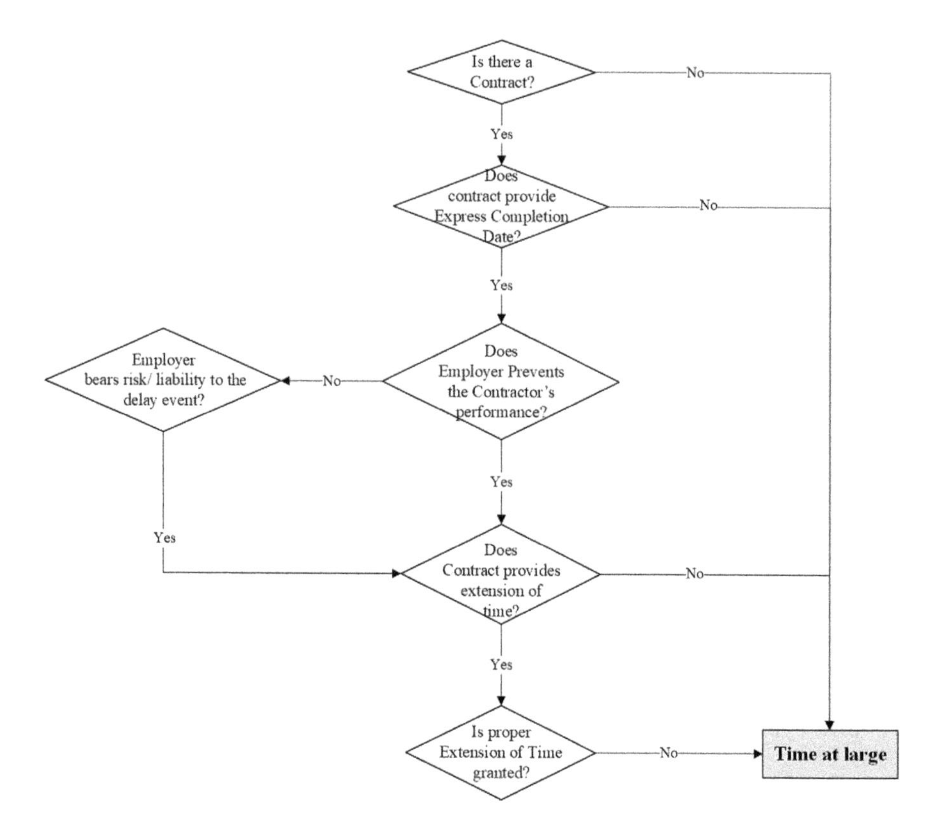

FIGURE 3.13 Situations for Time at Large.

When the time is set to be at large, the Contractor can complete the works in a reasonable time. The question of what is reasonable time is determined under English common law. In *British Steel Corporation v. Cleveland Bridge and Engineering Co. Ltd* [1984] 1 All ER 504, the court provides that:

> *the question what constituted a reasonable time had to be considered in relation to the circumstances which existed at the time when the contractual services were performed but excluding circumstances which were under the control of the party performing those services.*

Time at large is an implied obligation to the Contractor in the absence of express contract completion date, but it is very difficult to verify in practice. Therefore, in order to avoid the uncertainty arising from the reasonable time, the Employer should clearly set out a fixed completion date in the contract.

3.5.3 Concurrent Delay

Concurrent delay is defined as "*a period of project overrun which is caused by two or more effective causes of delay of approximately equal causative potency*" (John Marrin QC, 2002).

Under the SCL Delay and Disruption Protocol second edition (2017), the true concurrent delay requires "*two or more delay events occur at the same time, one an Employer Risk Event, and the other a Contractor Risk Event, and the effects of these events on the project are felt at the same time.*" However, it is rare for two or more delay events to occur simultaneously in practice. Therefore, the industry usually takes a broader view of concurrent delay as two or more delay events that may occur in parallel in a period during which both may impact the completion date.

Concurrent delay is a complex area in construction disputes. Different approaches have been adopted in various cases in common law jurisdictions. Meanwhile, both the SCL's Delay and Dispute Protocol and the AACE's Forensic Schedule Analysis guideline provide the recommendations to deal with concurrent delay. However, none of these sources can provide a clear conclusion of where the dispute relating to concurrent delay would be determined.

3.5.3.1 English Law – Malmaison Approach

Under English law, the initial principle applied for concurrent delay is set out in *Henry Boot Construction v Malmaison Hotel* (1999) 70 Con LR 33. In this case, the court adopted a dominant approach, which used the "But For" test to evaluate which party is actually responsible for the dominant cause of the delay. This approach is also known as the Malmaison approach. In general principle, in the event of concurrent delay, the Employer and the Contractor share the risks, in which the Contractor gets the extension of time, but no compensation for the associated delays.

For example, in *Walter Lilly & Company Ltd v Mackay and another* [2012] EWHC 1773, the court followed the Malmaison approach and concluded that if the delay was caused by two or more events, one of which entitles the Contractor to the extension of time, then the Contractor is entitled to the full extension of time.

In line with the Malmaison approach, the Society of Construction Law Delay and Disruption Protocol (2017) provides the guideline to deal with the concurrent delay as follows:

> *True concurrent delay is the occurrence of two or more delay events at the same time, one an Employer Risk Event, the other a Contractor Risk Event, and the effects of which are felt at the same time. For concurrent delay to exist, each of the Employer Risk Event and the Contractor Risk Event must be an effective cause of Delay and Completion (i.e. the delays must both affect the critical path). Where Contractor Delay to Completion occurs or has an effect concurrently with the Employer Delay to Completion, the Contractor's concurrent delay should not reduce any EOT due.*

Furthermore, in *North Midland Building Limited v Clyden Homes Limited* [2018] EWCA Civ 1744, because NMBL intended to levy the liquidated damages for periods of concurrent delay, the English Court of Appeal held that the Employer should not benefit from the delay that they are also responsible for.

3.5.3.2 Scottish Law – Appointment Approach

In Scotland, a rule for the appointment approach for concurrent delay arose in *City Inn v Shepherd Construction* (2007) COSH 190. With the appointment approach, in the event of two or more causes for concurrent delay of the project, if none of the causes can be identified as dominant, the Contractor and the Employer will share the delay proportionately as determined by their associated risks.

3.5.3.3 US Law – Multiple Approaches

In the United States, the US courts apply three approaches when considering the concurrent delay:

1) the Contractor is entitled to the extension of time but no cost compensation
2) Responsibility based on critical path analysis
3) Appointment approach

The approaches that the court of the United States adopted for the concurrent delay is specified in the AACE's Forensic Schedule Analysis guidelines, which rules to deal with the concurrent delay by considering both the responsibility of the party causing the delay and the contractual risk allocation of the delay event. It defines that the Contractor's or Employer's delay as "*any delay event caused by the Contractor or Employer, or the risk of which has been assigned solely to the Contractor or Employer*" (AACE, 2012). If the Contractor's delay is concurrent with the Employer's delay or is a force majeure delay, then the Contractor is entitled to the extension of time but no monetary compensation. If the Employer's delay is concurrent with a force majeure delay, then the Contractor is also entitled to the extension of time but no monetary compensation. However, if the Employer's delay is concurrent with another Employer's delay or nothing, then the Contractor is entitled to both the extension of time and monetary compensation.

In the absence of contract provisions, in the event of concurrent delay, both parties are impacted with their rights to claim damages. The Contractor is barred from delay damages and the Employer is barred from liquidated damages, to offset the delays caused by the other party.

If the Contractor's delay is on the critical path, excluding concurrent causes of a force majeure event, the Contractor is neither entitled to extension of time nor delay compensation. In contrast, if the Employer's delay is on the critical path, excluding concurrent causes of a force majeure event, the Contractor is entitled to extension of time and compensation related to the delay. If the delay is partially caused by force majeure event, then the Contractor is entitled to extension of time but no delay-related compensation. The claimant has the burden of proof to the appointment.

Further, in *Coath & Goss Inv v US* 101 Ct Cl 702 (1944), the US Court of Claims applied the appointment approach and held that:

> *where both parties contribute to a delay neither can recover damage unless there is in the proof a clear appointment of the delay and the expense attributable to each party.*

3.5.3.4 Australia Law

In Australia, Clause 34.4 of standard form AS 4000 provides:

> *when both non-qualifying and qualifying causes of delay overlap, the Superintendent shall apportion the resulting delay to WUC according to the respective causes' contribution.*

> *In assessing each EOT the Superintendent shall disregard questions of whether:*
>
> *a) WUC can nevertheless reach practical completion without an EOT; or*
> *b) the Contractor can accelerate,*
>
> *but shall have regard to what prevention and mitigation of the delay has not been effected by the Contractor.*

Further, Clause 35.3 of standard form AS 2124-1992 provides:

> *where more than one event causes concurrent delays and the cause of at least one of those events but not all of them, is not a cause referred to in the preceding paragraph, then to the extent that the delays are concurrent, the Contractor shall not be entitled to an extension of time for Practical Completion.*

In FIDIC 2017 contract, the last paragraph of Sub-Clause 8.5 mentions concurrent delay as follows:

> *If a delay caused by a matter which is the Employer's responsibility is concurrent with a delay caused by a matter which is the Contractor's responsibility, the Contractor's entitlement to EOT shall be assessed in accordance with the rules and procedures stated in the Special Provisions (if not stated, as appropriate taking due regard of all relevant circumstances).*

The assessment of a concurrent delay should first rely on the agreement between the parties. Therefore, it is important for the parties to agree and record the ownership of risks in the event of concurrent delay and establish rules for granting the extension of time. In *North Midland Building Ltd v Cyden Homes Ltd* (CA) [2018] EWCA Civ 1744, Clause 2.25 of the JCT contract has been amended as "*any delay caused by a Relevant Event which is concurrent with another delay for which the Contractor is responsible shall not to be taken into account.*" The court considered that parties agreed that the Contractor bears risk of concurrent delay in the express terms, which overrides the prevention principle, which would otherwise be implemented under common law. The guidance note of FIDIC 2017 contract also suggests that the parties apply the concurrent delay rules set out by the Society of Construction Law Delay and Disruption Protocol in the contract-specific conditions.

The rule set out by the AACE makes the critical path more important. In reality, when investigating the delay event in sufficient detail, rarely can a true concurrent delay be found. In order to avoid uncertainty associated with the concurrent delay claims, it is better to develop the programme in sufficient detail. In conjunction with the PRINCE2 (Axelos, 2017) planning principles, it is difficult to provide a detailed programme from the beginning of large construction projects. It is worth considering agreeing at the high-level stage programme and agreeing the milestone and/or key dates for each stage, then the Contractor can develop the detailed next stage programme toward the end of the previous stage. This approach allows a more realistic and accurate baseline programme established for the project and the programme can then be better used to monitor and control the project as well as to undertake delay analysis.

NEC is silent on the concurrent delays, as the Contractor is entitled to extension of time under all compensation events set out in the contract no matter whether it is concurrent with its own delay. Unlike JCT, NEC intends to take a proactive approach to manage changes and risks as the project progresses; the contractual completion date is adjusted with the time compensation for every implemented compensation event and no retrospective review and changes are allowed for the implemented compensation event in accordance with Clause 66.2. Therefore, in the event of concurrent delay, the Contractor is still entitled to the delay arising from the compensation event; it may not be the same as what was delayed by the Contractor themselves, but the Contractor's rights stand nevertheless, although the background changes. NEC clearly sets out the Employer's risks and the Contractor bears the remaining risks. Therefore, it is reasonable that the Contractor is entitled to all the compensation arising from the compensation event. This approach is also in line with the "Impacted As-Planned Analysis" and "Collapsed But For Analysis" methods. It is a simpler and reasonable way to deal with the difficulties that arise in the concurrent delay claims.

3.5.4 Liquidated Damages

Liquidated damages are the pre-agreed remedy between the contract parties even if the Contractor fails to complete the works by completion date. Liquidated damages can be classified as delay liquidated damages in the event of failure to complete the works within a specified time, and performance liquidated damages for the Contractor's failure to reach the required level of performance. There is no need for the Employer to prove actual loss to claim the liquidated damages.

On the one hand, the Contractor will pay the Employer for any time it delayed from the completion date; on the other hand, it may incentivize the Contractor to complete the works on time due to the potential payment the Contractor is obligated to make after the contractual completion date. During the contract drafting, the Contractor should understand its risk on schedule when undertaking the project and agree a reasonable rate for liquidated damages with the Employer. In large construction projects, the liquidated damage is usually defined as a fixed daily or weekly rate. The total liquidated damage is calculated as the delay between the actual completion date and the contractual completion date, multiplied by the pre-agreed fixed weekly or daily rate. Some contracts also set out a cap for the total liquidated damages for the Contractor, such a cap is usually 10% to 20% of the whole contract value.

3.5.4.1 Common Law

The concept of liquidated damages can be traced back to *Astley V Frances Weldon* [1801] EngR 108. The long-standing position of testing the validity of liquidated damages is set out in *Dunlop Pneumatic Tyre Co Limited v. New Garage and Motor Co* [1915] AC 79, where the court provides:

> *the essence of a penalty is a payment of money stipulated as in terrorem of the offending party; the essence of liquidated damages is a genuine covenanted pre-estimate of damage.*

The judgment of "penalty" or "liquidated damages" is not based on the words used in the contract, but on finding out whether the payment is a true penalty or liquidated damages. Further, it will be judged at the time of the contract formation, thus the situation at the time of breach is irrelevant.

In paragraph 86 of the judgment, Lord Dunedin set out four rules to test whether it is an unenforceable penalty or enforceable liquidated damages as follows:

1) The amounts payable is "*extravagant and unconscionable in amount in comparison with the greatest loss that could conceivably be proved to have followed from the breach*".
2) "*if the breach consists only in not paying a sum of money, and the sum stipulated is a sum greater than the sum which ought to have been paid*".
3) In the event of "*a single lump sum is made payable by way of compensation, on the occurrence of one or more or all of several events, some of which may occasion serious and others but trifling damage*," a presumption of penalty will be made.
4) The presumption made above would not impact the decision to liquidated damages if a genuine pre-estimate of damage is found.

However, in *Parkingeye Ltd v Beavis and Cavendish Square Holding BV v Talal El Makdessi* [2015] UKSC 67, the English Supreme Court re-formulated the test of penalty clause from "genuine pre-estimate of the loss" to "legitimate interests." The Court states that the penalty rule is "*an ancient, haphazardly constructed edifice which has not weathered well*," but only effected a modest refinement of the position of English law. The key issue of whether a clause is a genuine pre-estimate of loss may be helpful for straightforward liquidated damages' clauses, but is insufficient for more complex situations. The real test as to whether the clause is a penalty is whether the provision goes beyond the relevant party's legitimate interest.

In *Parkingeye v Beavis*, Beavis parked in a car park where the first two hours were free and a fine would be applied afterwards. As a consequence of overstay, Beavis was fined £85, but he argued that the fine was a penalty, which is unenforceable. The English Supreme Court held that:

> *deterrence is not penal if there is a legitimate interest in influencing the conduct of the contracting party which is not satisfied by the mere right to recover damages for breach of contract.*

In *Cavendish Square v Makdessi*, Makdessi sold his 60% of shares of his consulting company to Cavendish Square in instalment payments. The contract provides that if Makdessi breaches certain restrictive covenants, Makdessi would not be entitled to the interim and final payments and Cavendish Square can buy Makdessi's remaining shares at net value. The Supreme Court ruled the new penalty test as follows:

> *whether the impugned provision is a secondary obligation which imposes a detriment on the Contractor breaker out of all proportion to any legitimate interest of the innocent party in the enforcement of the primary obligation.*

Therefore, post *Cavendish Square v Makdessi*, the court no longer considers whether the amount set out for the liquidated damages is a genuine pre-estimate of the loss likely to be suffered, or whether the actual loss is significantly less than the predetermined sum. Once the parties agree a specific rate of the liquidated damages, it is bound to both parties.

Different from the English Court approach, in the United States, the court determines whether the liquidated damages clause is enforceable based on two criteria as follows:

1) uncertainty: whether the damages arising from the breach are difficult to calculate
2) reasonableness: whether the amount due is reasonable compared with the actual/anticipated damages

3.5.4.2 Civil Law

In the civil law system, traditionally there is no distinction between the liquidated damages and penalty. In fact, many civil law codes usually allow penalty in order to incentivize the performance, e.g., Article 114 of the Chinese Contract Law, Article 1154 of the Spanish Código Civil, and Articles 1226 to 1233 of the French La Code civil.

In civil law jurisdictions, the tribunal may adjust liquidated damages provisions to reflect more accurately the actual position between the parties in the event of a breach. For example, Articles 1152 of the France La Code civil, "*liquidazione convenzionale del danno*" of the Italian Civil Code, and Articles 340–341 of the German BGB.

3.5.4.3 Standard Forms Contract

Because the liquidated damages are at a pre-agreed fixed rate, it provides certainty for both parties in the event of delay and it saves time and the expense of arguing over the quantity of general damages. Under English law, the Employer does not need to prove an actual loss caused by the delay, but rather only prove that the Contractor is responsible for the delay. Consequently, most construction contracts provide Liquidated Damages provisions, e.g., Sub-Clause 8.8 of FIDIC 2017 Red Book, Option X7 of NEC4, and Clause 2.34 of JCT Standard Building Contract 2016.

3.5.4.4 FIDIC

Sub-Clause 8.8 "Delay Damages" of FIDIC 2017 Red Book provides:

> *If the Contractor fails to comply with Sub-Clause 8.2 [Time for Completion], the Employer shall be entitled subject to Sub-Clause 20.2 [Claims For Payment and/ or EOT] to payment of Delay Damages by the Contractor for this default. Delay Damages shall be the amount stated in the Contract Data, which shall be paid for every day which shall elapse between the relevant Time for Completion and the relevant Date of Completion of the Works or Section. The total amount due under this Sub-Clause shall not exceed the maximum amount of Delay Damages (if any) stated in the Contract Data.*
>
> *These Delay Damages shall be the only damages due from the Contractor for the Contractor's failure to comply with Sub-Clause 8.2 [Time for Completion], other than in*

the event of termination under Sub-Clause 15.2 [Termination for Contractor's Default] before completion of the Works. These Delay Damages shall not relieve the Contractor from the obligation to complete the Works, or from any other duties, obligations or responsibilities which the Contractor may have under or in connection with the Contract.

This Sub-Clause shall not limit the Contractor's liability for Delay Damages in any case of fraud, gross negligence, deliberate default or reckless misconduct by the Contractor.

3.5.4.5 NEC

Option X7 "Delay damages" of the NEC4 contract provides:

X7.1 The Contractor pays delay damages at the rate stated in the Contract Data from the

- *Completion Date for each day until the earlier of*
- *Completion and*

the date on which the Client takes over the works.

X7.2 If the Completion Date is changed to a later date after delay damages have been paid, the Client repays the overpayment of damages with interest. Interest is assessed from the date of payment to the date of repayment.

X7.3 If the Client takes over a part of the works before Completion, the delay damages are reduced from the date on which the part is taken over. The Project Manager assesses the benefit to the Client of taking over the part of the works as a proportion of the benefit to the Client of taking over the whole of the works not previously taken over. The delay damages are reduced in this proportion.

3.5.4.6 JCT

Clause 2.34 "Liquidated Damages – Relevant Part" of JCT DB 2016 provides:

As from the Relevant Date, the rate of liquidated damages stated in the Contract Particulars in respect of the Works or Section containing the Relevant Part shall reduce by the same proportion as the value of the Relevant Part bears to the Contract Sum or to the relevant section Sum. As shown in the Contract Particulars.

If the Contractor fails to complete the works by the contract completion date, the Employer should issue a "Non-Completion Notice" under Clause 2.28. Such a notification is a condition precedent for the Employer to deduct liquidated damages under Clause 2.29.1.1.

The specific rate of liquidated damages is usually provided in the Contract Data (NEC/FIDIC) or Contract Particulars (JCT). It is important to insert an appropriate rate in the field of liquidated damages. Errors made in this field may lead to a significant commercial impact on the contract parties. In *Temloc Ltd v Errill Properties Ltd* (1987) 39 BLR 30, the parties inserted "£Nil" in the contract particulars of the JCT contract; the court held that the Employer can not only claim the liquidated damages but also the general damages, because the contract contained a valid enforceable liquidated damages provision which is £0, hence no entitlement for a general damages claim. However, if the section of "liquidated damages" has been left blank instead, then the Employer is still entitled to claim general damages for the Contractor's breach of the contract.

However, the express contractual provisions for liquidated damages cannot be relied upon under the following circumstances:

a) in the circumstances of ambiguity or inconsistency, the contract will be interpreted under contra proferentum rule, e.g., in *Peak v McKinney*
b) the liquidated damaged is set up on a "penal" nature, e.g., *Dunlop v New Garage*
c) failure to comply with contractual procedures
d) in the circumstances that the time has been set at large, e.g., *City Inn v Shepard*
e) failure to comply with relevant statutory requirements
f) waiver or estoppel

For example, in *City Inn v Shepard*, the Contractor fails to meet the condition precedent, but it does not set time at large and liquidated damages remain claimable even though the Employer prevented the Contractor's performance.

3.5.5 Float Ownership

The planning techniques used in most large international construction projects is based on the Precedence Diagramming Method (PDM) which was developed in operations research for control of a defence project in the 1950s, as shown in Figure 3.14.

The start and finish date define the whole duration of the project, and there are usually multiple network paths with activities flowing from start to finish. Among all these paths, the longest path is the critical path of the project.

Float is the time built into a programme during which delays will not affect the Completion Date.

3.5.5.1 Who Owns the Float?

The ownership of the float can impact the entitlement of the actual extension of time and liquidated damages. In *Glenlion Construction v The Guinness Trust* (1987) 39 BLR 89, the judgment supports the argument that under English law the Employer owns the float rather than the Contractor.

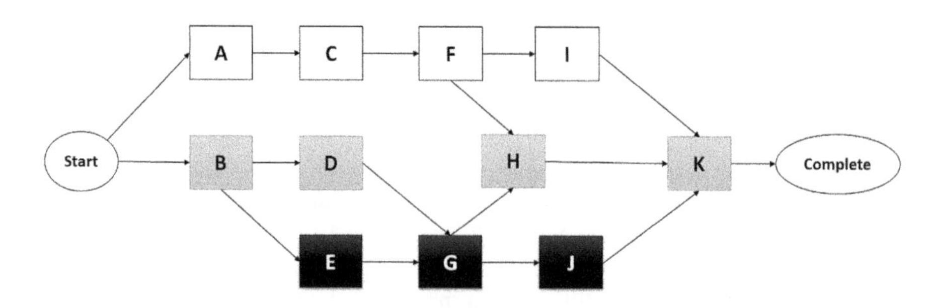

FIGURE 3.14 Precedence Diagramming Method.

Under Clause 31.2 of the NEC4 contract, the *Contractor* is obliged to show both the float and time risk allowance in the programme. There are three types of float under the NEC contract, namely Total Float, Time Risk Allowance, and Terminal Float.

Total float is the duration an activity can be delayed from its early start/finish date without delaying the completion date. It can be calculated as:

$$\text{Total Float} = \text{Late Finish} - \text{Early Finish} = \text{Late Start} - \text{Early Start}.$$

Equation 3.1 Calculation of Total Float

Time Risk Allowance is the amount of time contingency allowed by the Contractor in activities for potential risk of delay. This is usually allowed for the foreseeable common risks that have a relatively high probability of occurring and low severity impact, e.g., resource productivity, poor weather conditions, and inefficient interface collaboration, etc.

Terminal Float is the difference between the planned completion and contract completion dates, as shown in Figure 3.15.

Different types of float are associated with different entities in the programme and may derive different ownership. Table 3.4 describes the different floats and their features under the NEC contract.

The ownership of the float often arises as an issue when deciding the amount of extension of time.

The NEC contract clearly provides the owner of each type of float. The total float is shared between the *Client* and the *Contractor* on a first come, first served basis; whereas the Terminal Float and Time Risk Allowance are owned by the *Contractor*.

In the absence of the contract provisions, Section 8 of the SCL Delay and Disruption Protocol provides (SCL, 2017) that:

> *where there is remaining total float in the programme at the time of an Employer Risk Event, an EOT should only be granted to the extent that the Employer Delay is*

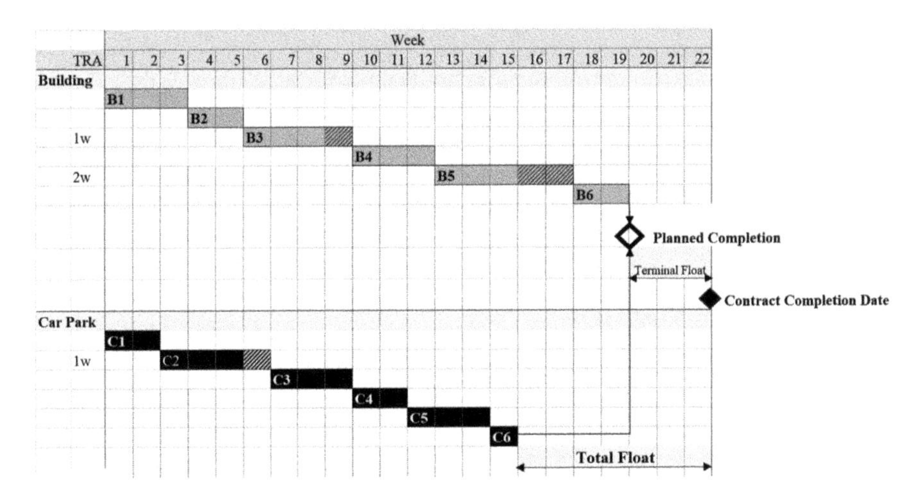

FIGURE 3.15 Float type under the NEC contract.

TABLE 3.4
Types of Float under NEC Contract

Float Type	Calculation	Associated to	Owner	Who can use
Total Float	Late Finish – Early Finish or Late Start – Early Start	Activity or Network Path	Shared between Contractor & Employer	First come, first served
Time Risk Allowance	Activity Duration Uncertainty	Each Activity	Contractor	Contractor
Terminal Float	Contract Completion Date – Planned Completion Date	Whole Project	Contractor	Contractor

predicted to reduce to below zero the total float on the critical path affected by the Employer Delay to Completion.

Therefore, the SCL's approach takes the progress to the point of when the Employer Delay occurs. If there is any total float left in the programme, the Employer will use the float and the Contractor's entitlement for the extension of time will be calculated by the duration of delay caused by the Employer minus the total float.

Therefore, the SCL's approach still follows the first come, first served principle. If the Contractor delays certain activities during the project implementation, it can use the float in the programme. When the Employer's delay events occur, the Employer can also use the float remains in the programme.

3.5.6 Time Bar and Condition Precedent

During the claim of extension of time, the parties shall pay attention to the potential time bar as a precedent condition for a valid claim. Clause 61.3 of NEC4 provides the time bar for the Contractor's entitlement to claim compensation event if the *Contractor* is obliged to provide the notification of the compensation event. Likewise, Clause 20.2.1 of the FIDIC 2017 contract requires notification within 28 days as the condition precedent for the Contractor to claim additional time or cost. In contrast, JCT does not apply time bar to the Contractor's notification of delay, and Clause 2.24 requires the Contractor to notify the Relevant Event within a reasonable time. Consequently, there are more delay claims and disputes toward the end of the project on a JCT contract. The detailed time bar issues will be discussed in Chapter 6, Change Management.

3.6 RECOMMENDATIONS

The detailed programme requirement constitutes one of the success factors of the NEC3 contract over the past decade. With the intention to incorporate global best practice, the FIDIC 2017 edition also provides detailed requirements for the programme as well as a detailed supporting report associated to the programme. This is a good move which will eventually lead to effective project control from the beginning of the project.

However, in order to improve the overall project planning and control of the programme, it is important to set out the standardized planning procedure at the beginning of the project. PMBOK of PMI (2017) sets "prepare the schedule management plan" as the first step for project schedule management. This plan will set out the standard and planning tool to be used, the standardized work breakdown structure (WBS), activity ID structure, activity code, the standardized calendars, the standardized user defined field, e.g., QSRA and TRA, the resource and cost loading principle, the schedule quality measurement criteria, and overall control and administration.

As the programme in large construction projects often has over 10,000 activities and multiple planners may work simultaneously on the same programme in the network or cloud-based platform and the same activity name may be used for similar works in different sections, it is difficult to tell if the logic has been correctly established during the programme review. It is good practice to standardize the activity ID in the planning procedure or project control procedure at the beginning of the project. The activity ID should set out standard code for the work package, the organization, the location or section in which the work is based, the stage of the work, or the discipline. For example, L2.WP3.S1.CV.010 represents the Lot 2, work package 3, stage 1, Civils' work. The planner can immediately identify where the predecessor or the successor linked with each activity is and it is useful when reviewing the relationships, tracing the critical path, and correcting loops during project planning.

Due to the complexity of the detailed programme of large construction projects, not everyone can access P6 and is able to read the complex logic among thousands of activities. It is recommended that each level of management provides a different level of the programme; although the P6 can simply collapse the detailed programme in each level of the work breakdown structure (WBS). In addition, it is useful to provide the one-page level one programme to demonstrate the overall process and milestones of the programme for communication with the Client and stakeholders.

In addition, it is worth establishing a template to export the period programme from P6 to allow the site manager to develop the daily work plan to manage and control the daily site works.

It is recommended that the planning manager or project control manager undertake regular schedule quality checks and facilitate the section planner to rectify the identified issues, e.g., open-ended logic link, the excessive lags or lead, the start to finish relationship, etc. The regular schedule checks will maintain good planning practice across the project which will then lead to a reliable programme to be used for analysis of the critical path, assessing changes and associated extension of time, to project the forecast, and control the project delivery.

If the resource is available, it is also beneficial to undertake regular quantitative schedule risk analysis (QSRA) to identify problematic areas. QSRA can be used for various purposes. It can be used to assess the potential impact of the change and facilitate the decision of extension of time; it can also be used for comparing the different options to mitigate a risk. It provides the specific confidence level for a given completion date for each milestone. In addition, the criticality report following QSRA can be used to review the critical path, and control sub-critical paths. It also recommends that the parties check the confidence level of the Completion Date set out in the contract and it would be good to achieve at least 50% confidence in the first Accepted Programme.

Although the resource loading is not required for all construction programme, by doing so, the practicality of the programme can be reviewed with the resource profile. The project management team can use it to optimize the construction sequence as well as the resource utilization. Therefore, the resource loading will be eventually be of benefit to the overall project planning and project control. Sub-Clause 8.3.(k).(iii) of the FIDIC 2017 contract requires the Contractor to provide a detailed estimation of its personnel and equipment. Although it is not mandatory for the Contractor to be required to resource load the programme, and the Contractor can provide the required information through other approaches, the resource loaded programme is the most appropriate approach to comply with the contract requirement. It provides an integrated platform to manage both the programme and cost at the same time in a dynamic manner.

Another feature of the NEC and FIDIC 2017 forms of contract is the dynamic baseline programme. As the latest Accepted Programme by the *Project Manager* under NEC or the Engineer under FIDIC 2017 will be used to assess the future compensation event or variations and claims, this creates an issue in relation to the float. As discussed in Section 3.3.2, NEC clearly specifies the ownership of float for each type of float, however FIDIC does not provide such clarification. When the progress is updated at each period, the activity dates are often changed period by period. Therefore, instead of comparing with the first Accepted Programme which is the traditional baseline programme, now both NEC and FIDIC measure the delay based on the latest Accepted Programme, which can be quite different from the first programme as time goes by. This means that the general float is taken based on the first come, first served principle. When the float is reduced, the change of critical path may become frequent from period to period and toward the end of the project, the average criticality of the remaining activities is expected to increase. Consequently, the changes toward the end of the project are more likely to trigger extension of time hence the increase in the associated cost. The Employer should take this into account when managing the programme from the beginning of the project and establish appropriate control at the front-end of the project.

REFERENCES

Book/Article

AACE. 2011. *Forensic Schedule Analysis, TCM Framework: 6.4 – Forensic Programme Assessment.* AACE International Recommended Practice No. 29R-03. Morgantown, WV: AACE International Inc.

Axelos. 2017. *Managing Successful Projects with PRICE2.* 2017 Edition. Norfolk, UK: TSO (The Stationery Office) / Williams Lea Tag.

Defense Contract Management Agency (DCMA). 2012. *Earned Value Management System (EVMS) Programme Analysis Pamphlet (PAP).* https://www.dcma.mil/Portals/31/Documents/Policy/DCMA-PAM-200-1.pdf?ver=2016-12-28-125801-627, Accessed on 11 November 2019.

Halsbury's Laws of England. 2019. *Volume 6: Building Contracts*. 5th Edition. London, UK: LexisNexis.
PMI. 2017. *A Guide to the Project Management Body of Knowledge, PMBOK Guide*. 6th Edition. Newtown Square, PA: Project Management Institute.
SCL. 2017. *Delay and Disruption Protocol*. 2nd Edition. London, UK: The Society of Construction Law.

Contract

FIDIC. 1999. *Conditions of Contract for Construction*. 1st Edition. (1999 Red Book). Geneva, Switzerland: The Fédération Internationale des Ingénieurs-Conseils.
FIDIC. 2017. *Conditions of Contract for Construction*. 2nd Edition. (2017 Red Book). Geneva, Switzerland: The Fédération Internationale des Ingénieurs-Conseils.
FIDIC. 2017. *Conditions of Contract for EPC / Turnkey Project*. 2nd Edition. (2017 Yellow Book). Geneva, Switzerland: The Fédération Internationale des Ingénieurs-Conseils.
FIDIC. 2017. *Conditions of Contract for Plant & Design Build*. 2nd Edition. (2017 Silver Book). Geneva, Switzerland: The Fédération Internationale des Ingénieurs-Conseils.
JCT. 2016. *Design and Build Contract 2016*. London, UK: The Joint Contracts Tribunal Limited.
JCT. 2016. *Standard Building Contract with Quantities 2016*. London, UK: The Joint Contracts Tribunal Limited.
Mosey, David. 2014. *PPC2000 Standard Form of Contract for Project Patterning (Amended 2013)*. London, UK: The Association of Consultant Architects Ltd (ACA) and Association for Consultancy and Engineering.
NEC. 2013. *NEC3 Engineering and Construction Contract*. London, UK: Thomas Telford Ltd.
NEC. 2017. *NEC4 Engineering and Construction Contract*. London, UK: Thomas Telford Ltd.

Statutes

Consumer Rights Act 2015.
Supply of Goods and Services Act 1982.

Cases

Ascon Contractors Limited v. Alfred McAlpine Construction Isle of Man Limited (1997 ORB-361 & 1998 ORB-315).
Astley V Frances Weldon [1801] EngR 108.
Balfour Beatty Construction Ltd. v. The Mayor and Burgesses of the London Borough of Lambeth [2002] EWHC 597 [TCC].
City Inn v Shepherd Construction (2007) COSH 190.
Coath & Goss Inv v US 101 Ct Cl 702 (1944).
Dodd v Churton [1897] 1 Q.B. 562.
Dunlop Pneumatic Tyre Co Limited v. New Garage and Motor Co [1915] AC 79.
Hick v Raymond and Reid [1893] A.C. 22.
Holme v Guppy (1838) 150 E.R. 1195.
Multiplex Constructions (UK) Ltd v Honeywell Control Systems Ltd [2007] EWHC 447 (TCC).
North Midland Building Ltd v Cyden Homes Ltd (CA) [2018] EWCA Civ 1744.
North Midland Building Ltd v Cyden Homes Ltd [2018] EWCA Civ 1744.
Parkingeye Ltd v Beavis and Cavendish Square Holding BV v Talal El Makdessi [2015] UKSC 67.
Peak Construction (Liverpool) Limited v McKinney Foundations [1970] 1 BLR 111.

Shawton Engineering Ltd v DGP International Ltd [2005] EWCA Civ 1359.
Temloc Ltd v Errill Properties Ltd (1987) 39 BLR 30.
Tort British Steel Corporation v. Cleveland Bridge and Engineering Co. Ltd [1984] 1 All ER 504.
Trollope & Colls v North West Metropolitan Regional Hospital Board [1973] 1 WLR 601.
Union Eagle Ltd v Golden Achievement Ltd [1997] UKPC 5.
United Scientific Holdings Ltd v *Burnley Council* [1978] A.C. 904.
Wells v Army & Navy Cooperative Society (1902) 86 LT 764.

4 Cost Management

4.1 INTRODUCTION

Cost management is one of the most important aspects of large-scale construction project management. Due to the low profit margin for the construction industry, effective cost management not only increases the profit margin of the construction company, but also provides sufficient cash flow to support the project implementation. If the contractor, sub-contractor, or supplier make significant losses in a large construction project, it may lead to bankruptcy of the company and consequently may have a significant impact on the project. In addition, most disputes around construction projects are ultimately related to the payment between the parties. Therefore, effective front-end cost management can not only improve the construction company's profits, but also avoid disputes in the later stages of the project. This chapter begins by introducing the obligation between the contract parties in terms of project price and payment. It then explains the pricing and payment mechanism under the standard forms contract, including FIDIC, NEC, and JCT. Finally, it summarizes the best practice for effective cost control and emphasizes the pitfalls during cost management.

4.2 PRICE AND PAYMENT

Price and payment are the most important aspects in construction projects. The price is the value of the project agreed by the contract parties. It determines the Contractor's entitlement to payment by completion of the project, the payment procedures in each period during the project implementation, and final payment at the end of the project. It also determines the Contractor's cash flow as well as the final profit of the project undertaken.

4.2.1 Type of Contract

Contractual price and payment may proceed differently under different types of contract. Traditionally, the most commonly used contract is the lump sum or fixed price contract and the remeasurement contract. In addition, as a consequence of the success of the NEC contract, the target cost contract is often used in major infrastructure projects in the UK. Furthermore, the parties may also use other types of contract, e.g., cost plus, schedule of rate, or management fee.

4.2.1.1 Lump Sum

A lump sum contract is also known as a "fixed price" contract. The Contractor and the Employer agree the fixed price for the Contractor to complete the project. The price is generally estimated by the Contractor during the tender stage, which may involve some assumptions that the Contractor made in its estimation. The Contractor bears the risk of changing the assumptions during its estimation, unless it constitutes a change, which can then be claimed from the Employer. The Contractor bears more cost risk in this type of contract. Such contract will incentivize the Contractor to undertake thorough cost control as well as waste reduction action, e.g., through lean construction to increase profits. On the other hand, the Contractor normally includes a high percentage contingency to cover the potential risks it is responsible for. In a lump sum contract, the Contractor bears the risks of quantities. For example, in *Sharpe v San Paulo Railway* (1873) LR 8 Ch App 597, the Contractor had to bear the cost of the additional quantities even though it was doubled from that estimated in the tender.

4.2.1.2 Reimbursement

In contrast to the lump sum contract, in the reimbursement contract the Contractor gets paid for the actual cost it incurred during its performance. The Employer bears more cost risk in this type of contract. This type of contract is usually used in the project or a section of the project where the scope of the work is not clear enough, or in the professional service's agreement which will be largely engaged throughout the project on an on-going basis. In a remeasurement contract, the Employer bears the risk of any increase or decrease in quantities.

Further to completion of the works, or through progressive interim payments, the Contractor will be paid in accordance with the provisional quantities and this will impact the contract price.

4.2.1.3 Target Cost

The target cost contract usually pre-determines a price between the contract parties. The Contractor will be paid in accordance with the actual cost plus agreed profit margin. However, if the Contractor's final actual cost is different from the target cost, then the parties usually will share the savings and losses. However, in order to control the risk, the Employer often also sets up an upper lever threshold to cap the maximum overspend and then shares with the Contractor. Beyond the upper level threshold, the further loss will be within the Contractor's responsibility. Apart from NEC Option C and Option D, the target cost contract is also well-used in the framework or alliance contracts.

4.2.2 Statutory Obligation

In order to resolve the cash flow of the supply chain in the construction industry, Part II of the Housing Grants Construction and Regeneration Act (HGCRA) 1996 sets out the statutory payment requirement for construction projects under Sections 109 to 113. Part 8 of the Local Democracy, Economic Development and Construction Act (LDEDCA) 2009 then amended payment requirement under HGCRA and came into force from October 2011.

As shown in Figure 4.1, under the LDEDCA, the Contractor needs to submit the Application for Payment (AfP) before the due date, then the Employer needs to issue a payment notice specifying the notified sum within five days. If the Employer intends to pay less than the notified sum, they should serve a Pay Less Notice (PLN)

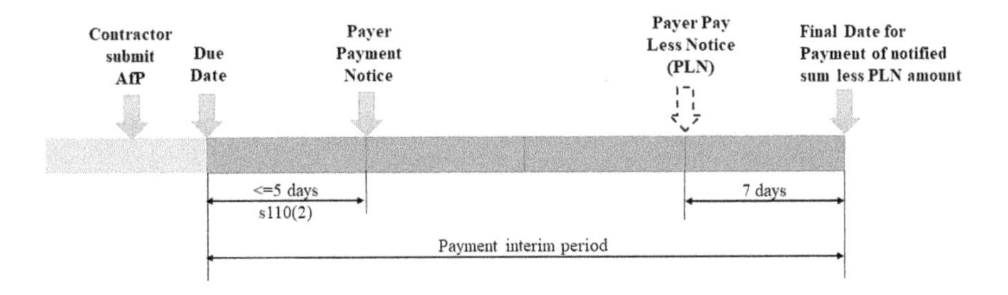

FIGURE 4.1 Payment requirement under Part 8 of the LDEDCA 2009.

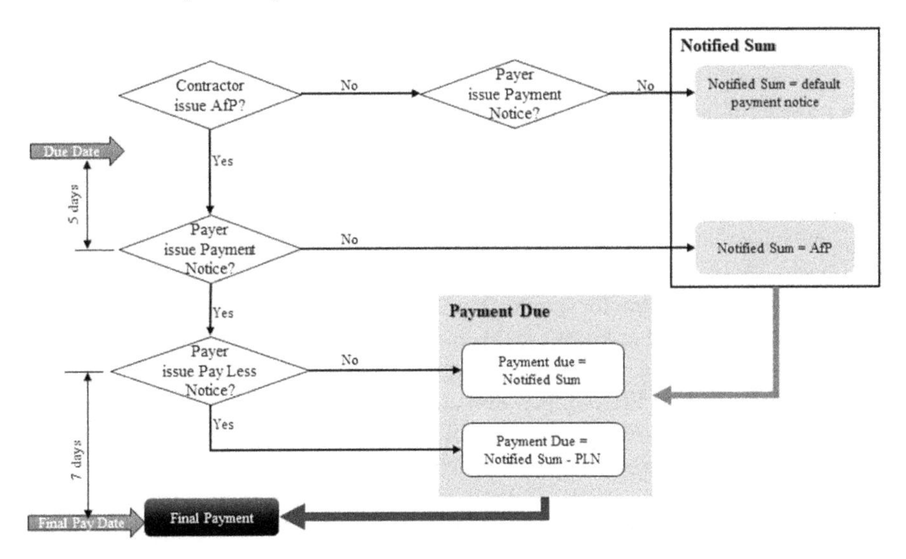

FIGURE 4.2 Default payment position under Part 8 of the LDEDCA 2009.

seven days before the final payment date. Then the payment due will be calculated as the notified sum minus the amount stated in the Pay Less Notice.

Figure 4.2 illustrates how the default payment position will be in case any party falls out in the payment process under Part 8 of LDEDCA 2009. If the Employer fails to issue the payment notice within five days of the due date, then the Contractor's application for payment will be set as the notified sum. If the Contractor fails to submit an application for payment, then the Employer should undertake their own assessment. If the Employer also fails to issue a payment notified, then the default payment notice will set out the notified sum. Likewise, if the Employer fails to issue the Pay Less Notice seven days before the final pay date, then the payment due will be the notified sum.

4.2.3 Employer's Obligation

The primary obligation for the Employer is to make payment in accordance with the contract as well as the statutory requirements. Under HGCRA 1996, if the Employer fails to make the payment, the Contractor can suspend the work at any time and the

Employer has no right to change Contractor due to such suspension. Meanwhile, the Contractor remains the entitlement to the payment of the cost incurred during its suspension due to the Employer's failure to pay. In such circumstances, the Contractor needs to serve a notice seven days before its suspension and the Contractor has to re-start work after the outstanding payment is made.

Even in jurisdictions that have no specific statutory payment requirement, the Employer is obliged to make payment in accordance with the contract. Failure to pay in time amounts to breach of contract and the Contractor is entitled to terminate the contract.

4.2.4 Contractor's Obligation

The Contractor's obligation in terms of cost management is more related to its own management of its account and associated records. For example, the staff timesheet, and site work labor timesheet, records associated to the material, plant and equipment usage, the communication in relation to the variations or change of the scope, etc. The Contractor is also responsible for maintaining sufficient funds to support the implementation of the project. Furthermore, the Contractor should ensure relevant insurances are in place and up to date.

In addition, the Contractor should comply with the contract required procedure in the event of change, otherwise its rights to claim additional time and/or cost may be time barred under some contract forms.

Sub-Clause 13.2 of FIDIC 2017 Red and Yellow Book provides the value engineering provisions and the Contractor will obtain 50% of the consequential savings. Like FIDIC, NEC4 introduces a new provision to incentivize the Contractor's value engineering input. For example, under the fixed price Option A and Option B of the NEC4 contract, Clause 63.12 provides that the Contractor will be awarded a percentage of savings for the compensation event, which is raised from the value engineering that the Contractor proposed.

4.3 COST MANAGEMENT UNDER FIDIC CONTRACT

Following the publishing of the fourth edition of the FIDIC Red book in 1987, it was widely used in large international construction projects. This success has continued with the FIDIC 1999 Red Book in the early 2000s. NEC option C, FIDIC Red Book as well as the FIDIC Pink Book are all reimbursement contracts, in which the Employer bears more risk. In order to control the overall project cost, the Employer may transfer risks to the Contractor, which has resulted in the design and build contract becoming dominant in recent years. Therefore, the FIDIC Yellow Book is now the most popular form of contract in the FIDIC family. Apart from the Yellow Book, the Silver Book and the Gold Book are also lump sum contracts.

4.3.1 Adjust the Contract Price

Under the FIDIC 2017 Red Book, the actual Contract Price is determined in measurement and valuation under Clause 12. Sub-Clause 12.3 "Valuation of the Works" sets out the procedure for the Engineer to evaluate the value of the works. Generally, the Engineer will use the quantity of works completed multiplied by the rate specified in the Bills of Quantity in the Schedule. A new contract rate may be applied under three circumstances: an un-identified item, excessive change of quantity, and instruction as variation. Sub-Clause 12.3(b) provides the measurement for the excessive change of quantity, as follows:

i) the variance of quantity is more than 10% from the Bills of Quantity in the contract
ii) the total value of changed quantity is more than 0.01% of the Accepted Contract Amount
iii) the quantity changes result in the unit cost rate changing by more than 1%
iv) the item is not defined as a "fixed rate item" under other sections of the contract

Under all Forms of the FIDIC contract, the Contract Price can also be changed through instruction of variations under Sub-Clause 13.3.1. The variation procedure is set out in Sub-Clause 13.3, which will be explained in Chapter 6 in more detail.

Under the FIDIC Red Book, the price adjustment due to the variation will refer back to Clause 12, which provides the same payment measurement approach as of the main works unless there is no rate or appropriate rate specified in the contract in accordance with Sub-Clause 12.3(a).

Under the Yellow and the Silver Books, the valuation of variation will be based on Cost Plus Profit, which is the Contractor's actual incurred expenditure plus the profit percentage stated in the Contract Data. Under Sub-Clause 1.1.20 of the Yellow Book and Sub-Clause 1.1.17 of the Silver Book, the default profit rate is 5% in the absence of specification in the Contract Data.

4.3.2 Payment

Unlike NEC, the FIDIC Red Book remeasurement contract only assesses the completed works to date based on the Bills of Quantity (BoQ). Therefore, the final Contract Price is uncertain until the project is completed under the Red Book; whereas the Contractor is paid for the proportion of the Lump Sum contract price based on the total works it completed to date under the Yellow Book.

Nevertheless, Sub-Clause 14.1(c) under both the Red Book and the Yellow Book expressly states that the quantity set out in the Schedule of the contract does not conclude the actual payment, which is based on the actual quantity completed for the works. However, the Contractor will use the Schedule to estimate the account of contract during the tender stage and submit for acceptance under the FIDIC Red Book. The Accepted Contract Amount will be used as the benchmark to calculate the Advanced Payment and Performance Security.

4.3.2.1 Advanced Payment

The Employer may need to provide advanced payment if it is set out in the Contract Data, for example by an interest free loan to mobilize the Contractor. In the absence of specification in the Contract Data, the Employer must pay the Contractor within 21 days of receiving the Advance Payment Certificate under Sub-Clause 14.7.(a). On the other hand, the Contractor may be required to provide an Advanced Payment Guarantee or Performance Security at the beginning of the project.

4.3.2.2 Interim Payment

Clause 14 sets out the procedure for the interim payment as explained in Figure 4.3.

The Contractor must submit a Statement and associated supporting documents to the Engineer after the end of each payment period in accordance with the requirements set out in Sub-Clause 14.3, which includes the following:

i) the estimated value of work completed
ii) adjustment as a consequence of changes in Laws under Sub-Clause 13.6
iii) retention amount
iv) adjustment for the advance payment or repayment under Sub-Clause 14.62
v) adjustment as a consequence of the Engineer's agreement or determination under Sub-Clause 3.7
vi) amount to be added for Provisional Sums under Sub-Clause 13.3
vii) release of Retention Money under Sub-Clause 14.9
viii) deduction of temporary utilities provided by the Employer to the Contractor under Sub-Clause 4.19
ix) deduction from the total of previous certified payments

The Engineer needs to issue an IPC (Interim Payment Certificate) within 28 days of receiving the Contractor's submission under Sub-Clause 14.6.1. The IPC needs to specify the amount due and any adjustment due under Sub-Clause 3.5 "Agreement or Determination".

The Contract Data may specify the minimum amount of IPC. The Engineer may withhold an IPC before issuing the Taking-Over Certification under Sub-Clause 14.6.2. The Engineer may also make a correction to any of the previous certified payments under Sub-Clause 14.6.3.

If the Contractor believes any item that he is entitled to as regards payment is missing in the Engineer's IPC, he needs to give a Notice to the Engineer and follow

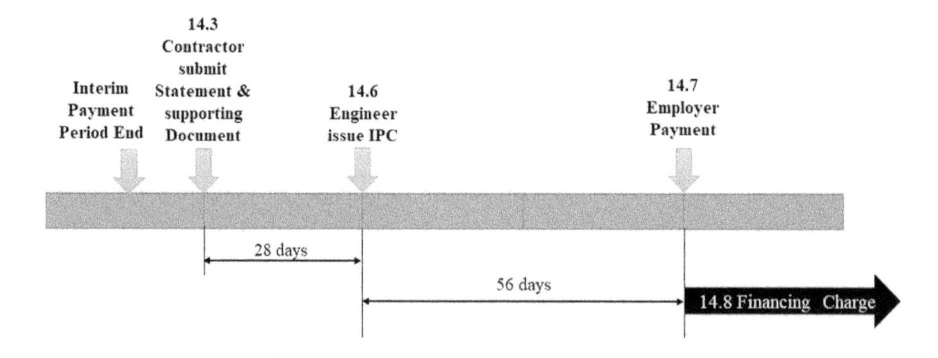

FIGURE 4.3 Payment process under FIDIC 2017 contract.

the procedure under Sub-Clause 3.7 to agree the amount, which will then be included in its next Application for the interim payment statement under Sub-Clause 14.3(vi).

In the absence of specification in the Contract Data, the Employer must pay the Contractor within 56 days of the Engineer issuing the IPC under Sub-Clause 14.7(b)(i).

If the Contractor does not receive payment within the time limit set out in Sub-Clause 14.7, the Contractor is entitled to the financial charges, which is calculated by a monthly compound rate to the outstanding payment under Sub-Clause 14.8. In the absence of specification in the Contract Data, the annual rate for the financial charge is 3%. The Contractor can request the payment of financial charge to which they are entitled without further application.

The total interim payment process generally takes 12 weeks under the FIDIC contract compared to the four-week payment cycle under the NEC contract. It is obvious that the cash flow position would be much better for the Contractor and relevant supply chains under the NEC contract, which realized the improvement as Sir Latham's intention.

Upon completion of the works, within 84 days, the Contractor shall submit the Statement with supporting documents as per requirements for interim payment application set out in Sub-Clause 14.3. The Engineer will then issue the IPC under Sub-Clause 14.6 and the Employer shall make payment in accordance with Sub-Clause 14.7.

4.3.2.3 Final Payment

When the defect liability period complete, within 56 days of issuing the Performance Certificate, the Contractor needs to submit a draft final Statement to the Engineer under Sub-Clause 14.11.1. The Contractor shall also submit the Statement of other amounts that they are entitled to post issuing of the Performance Certificate. Then the Contractor can submit the Final Statement to the Engineer. The Engineer then needs to issue the Final Payment Certificate (FPC) to the Employer within 28 days in accordance with Sub-Clause 14.13. In the absence of specification in the Contract Data, the Employer must pay the Contractor within 28 days of the Engineer issuing the FPC under Sub-Clause 14.7(b)(ii).

4.4 COST MANAGEMENT UNDER NEC CONTRACT

Unlike FIDIC which distinguishes the contract type in contract forms, NEC distinguishes the contract type through the main options. For all forms of NEC contract, Option A and Option B are lump sum contracts, and the remaining Option C, Option D, Option E, and Option F are remeasurement contracts. Different from FIDIC and JCT, NEC introduces Activity Schedule as a tool to assess the payment under Option A and Option C. Whereas, Option B, Option D, and Option E are all measured based on the Bills of Quantity. Option F is the management contract, which is measured based on the hourly rate of the time spent by the professional management consultant.

4.4.1 Payment

As the NEC contract developed as a consequence of the Latham Report, the payment provisions are set out in order to improve the overall cash flow across the supply chain of the construction industry. Payment under the NEC contract has some special features, in particular, Defined Cost and Disallowed Cost.

4.4.1.1 Payment Components

The primary source of payment is the Price for Work Done to Date (PWDD). The payment amount due for each interim payment is calculated by the PWDD plus other amounts to be paid to the *Contractor* under the contract, which then deducts the amount to be retained from the *Contractor* under the contract. The payment components under the NEC contract are described in Figure 4.4.

The PWDD is calculated differently under different main options. Under Option A, the PWDD is calculated with the Total of Prices for fully completed activities in accordance with Clause A11.2(29). Under Option B, the PWDD is the quantity of completed works at the rate of Bills of Quantity under Clause B11.2(30). Under Options C, D, E, and F the PWDD is the Defined Cost plus Fee in accordance with Clause 11.2(3). It worth noting that the Defined Cost under Options C, D, E, and F includes two components, the actual Defined Cost up to the current assessment date and the forecast Defined Cost to the next assessment date. In practice, because the *Contractor* is required to submit the Application for Payment (AfP) to the *Project Manager* before the assessment date, the AfP will include some *Contractor*'s forecast for the Defined Cost up to the next assessment date. Due to the complexity of large construction projects, this may involve hundreds of sub-consultants and thousands of staff and labor. It would take a couple of weeks to collect all information from all parties and prepare the consolidated application for payment. Therefore, it is more realistic to capture the accurate actual defined cost up to the previous assessment date and then include the forecast Define Cost for the current and next period, as shown in Figure 4.5.

The forecast Defined Cost is then corrected in the following application for payment when the actual Defined Cost is finalized.

4.4.1.2 Payment Schedule

NEC provides two different payments schedules depending on where the project is undertaken. For projects in the UK, the payment procedure needs to comply with the

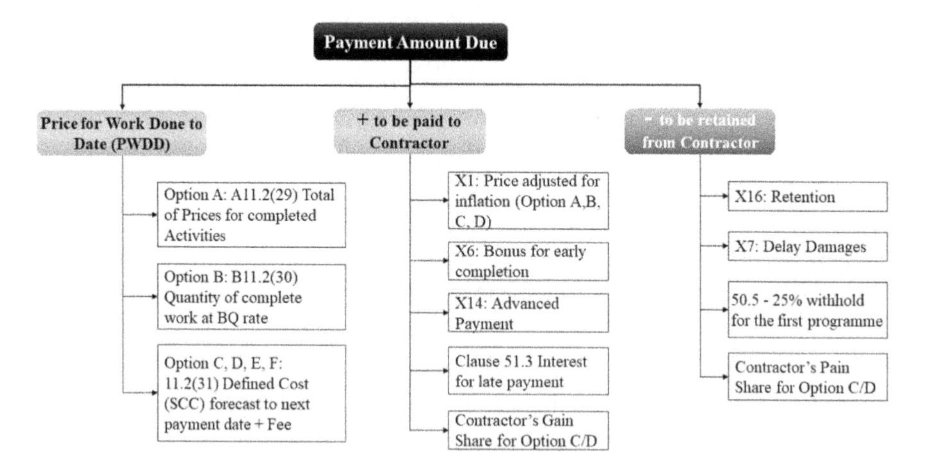

FIGURE 4.4 Payment component under NEC4 contract.

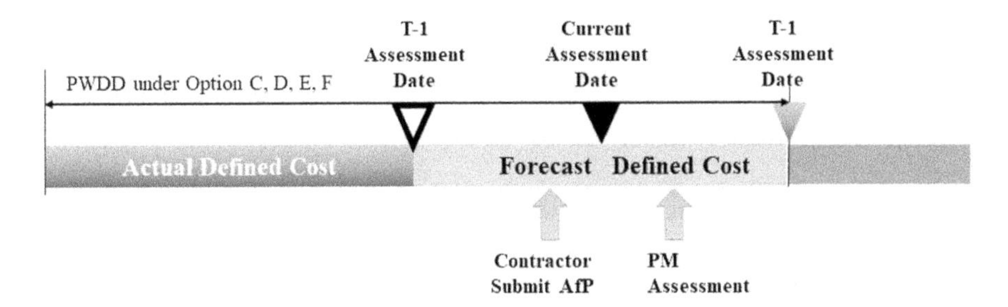

FIGURE 4.5 Defined Cost for monthly payment under Options C, D, E, F of NEC4 contract.

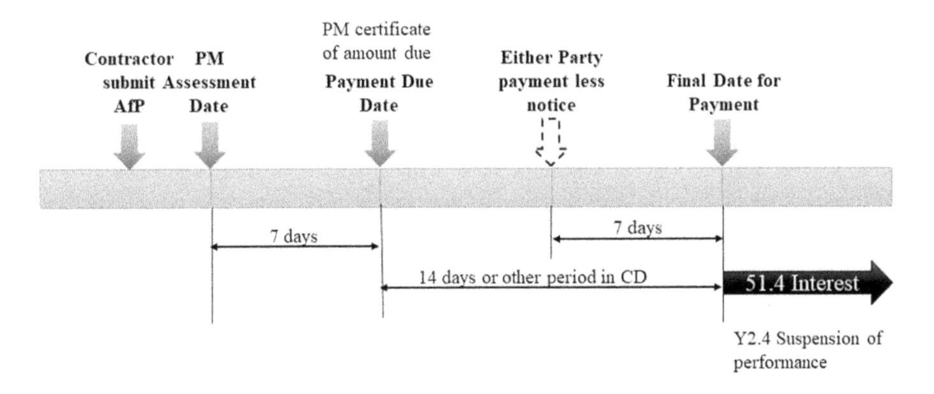

FIGURE 4.6 Statutory Payment Schedule under Option Y(UK)2.

statutory requirements set out in the Construction Act 2009. Figure 4.6 explains the payment process and time limit under the secondary Option Y(UK)2.

NEC4 requires the *Contractor* to submit the application for payment before the *Project Manager*'s assessment date. The *Project Manager* will assess the *Contractor*'s application for payment and serve a Payment Notice within seven days. The payment due date is then seven days after the *Project Manager*'s assessment date. The *Client* needs to pay the *Contractor* within 14 days of the payment due date. If the *Client* intends to pay less than the value stated in the payment notice, he needs to serve a Pay Less Notice no later than seven days before final date for payment.

Figure 4.7 demonstrates the Payment Schedule for a project outside the UK under Clause 51. The *Project Manager* needs to assess the *Contractor*'s application for payment within one week of each assessment date under Clause 51.1. Clause 51.2 then requires the *Client* to pay the *Contractor* the certified amount due within three weeks of the assessment date. Failure to make the payment on time will be subject to the interest calculated in the daily basis under Clause 51.4.

4.4.1.3 Payment Assessment

In NEC4, the new Clause 50.2 requires the *Contractor* to submit the Application for Payment (AfP) before each assessment date, in order to incentivize the *Contractor* to

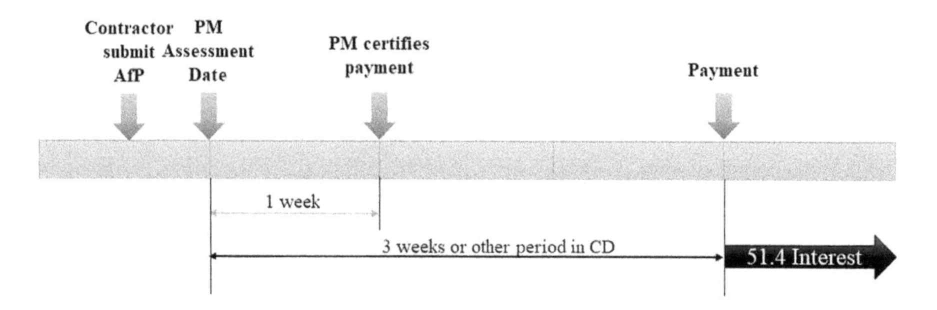

FIGURE 4.7 Payment Schedule outside the UK under Clause 51 of NEC4.

submit the application for payment. If the *Contractor* fails to submit the application for payment before the assessment date, then the *Project Manager* only assesses the amount due by the assessment date and the *Contractor*'s application for payment submitted before the assessment date under Clause 50.3. In contrast, if the *Contractor* submits the application for payment before the assessment date, then the *Project Manager* will assess the amount due as per the NEC3 approach, which includes "*the Price for Work Done to Date, plus other amounts to be paid to the Contractor, less amounts to be paid by or retained from the Contractor*" under Clause 50.4. Therefore, the *Contractor* will get paid less if they fail to submit the application for payment before the assessment date.

The actual payment will be assessed based on the Price for Work Done to Date (PWDD). The components of PWDD is different under different main options. Under Option A, the PWDD is the sum of the 100% completed activities to the assessment date. Under Option B, the PWDD is the total of the quantity of works completed to the assessment date. The PWDD is more complicated under Option C and Option D, which are target cost contracts. Figure 4.8 explains the components of the Price for Work Done to Date under NEC4 ECC Option C and Option D contracts.

The PWDD is calculated by the total Defined Cost plus the total Fee and then minus the total Disallowed Cost.

4.4.1.4 Final Payment

Under Clause 53.1, the *Project Manager* should provide his assessment of the final account to the *Contractor* within three weeks after either four weeks after the *Supervisor* issuing the Defects Certification or thirteen weeks after the *Project Manager* issuing a termination certification. If the *Project Manager* fails to submit his final account assessment within three weeks, then the *Contractor* can submit his assessment of the final amount and submit it to the *Client* in accordance with Clause 53.2. If the *Client* agrees with the assessment of the final account either by the *Project Manager* or the *Contractor*, the payment must be made within three weeks of the assessment date.

If the Parties have any disputes in regard to the final amount, it will refer to dispute resolution under Options W1, W2, or W3 in accordance with Clause 53.3. The final amount due will then be adjusted in accordance with the result concluded via the dispute resolution process under Clause 53.4.

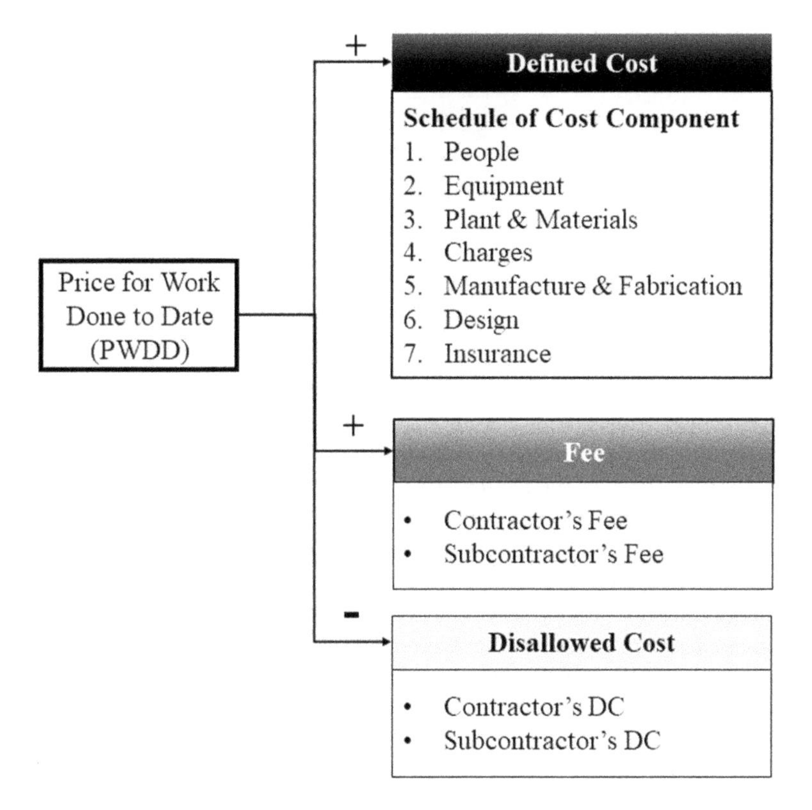

FIGURE 4.8 Components of the Price for Work Done to Date under NEC4 ECC Option C and D.

4.4.2 Defined Cost

Defined cost is important to the payment as well as assessing the compensation event under the NEC contract. Under all NEC main options, the compensation event is evaluated based on the Defined Cost. The defined cost has different definitions under each main option. For main Options C, D, E, and F, defined cost is the basis of the payment assessment, therefore it is important.

Defined Cost under the Schedule of Cost Components (SCC) and the Short Schedule of Cost Components (SSCC) include the following components:

1) People
2) Equipment
3) Plant and materials
4) Charges
5) Manufacture and fabrication
6) Design
7) Insurance

Unlike the main options, the payment approach is divided into different regimes; the compensation event under all main options is assessed by the Defined Cost plus Fee.

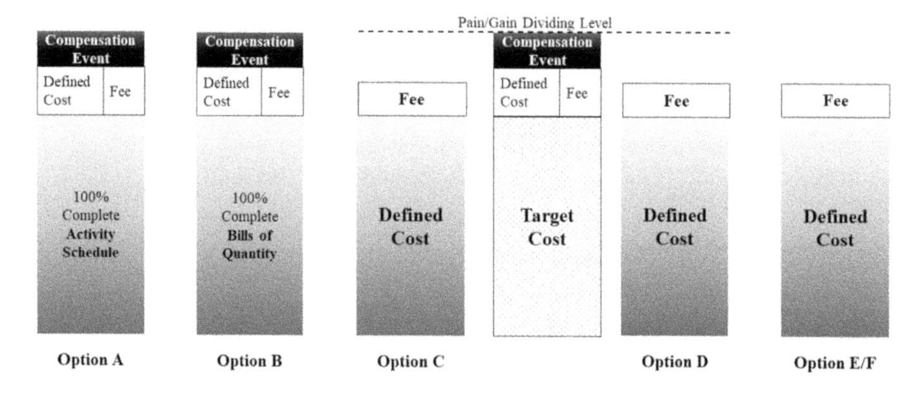

FIGURE 4.9 Defined Cost impact on the PWDD under NEC4 Main Options.

As a result, the impact of the Defined Cost to the Price of Work Done to Date under different main options can be different, as described in Figure 4.9.

Under Options A and B, the fixed price is defined in the contract. The payment will be assessed by the 100% completed works based on the Activity Schedule under Option A or Bills of Quality under Option B, plus the defined cost associated with the implemented compensation event and Fees.

Under Options C and D, the value of the compensation event is only used to define the revised target cost which will determine the value where the Pain/Gain share will be calculated. The actual PWDD is determined by the Defined Cost in accordance with the Schedule of Cost Components plus Fee.

As Option E and F is a cost reimbursable contract, although the compensation event can be used to forecast the project outturn, it does not have much impact on the payment, which is always based on the Defined Cost plus Fee.

Under NEC Options C, D, E, and F, the payment is based on the Defined Cost. Clause 50.9 of NEC4 sets out the procedure for the *Project Manager* to review the Defined Cost, as shown in Figure 4.10.

When the Defined Cost is finalized with supporting detailed records, the *Contractor* notifies the *Project Manager*. The *Project Manager* then needs to make a decision within 13 weeks, otherwise the *Contractor*'s Defined Cost is deemed to be accepted. If the *Project Manager* requires further information, the *Contractor* needs to provide additional records as requested by the *Project Manager* within four weeks, and the *Project Manager* needs to make his decision within four weeks. Likewise, the sanction on the *Project Manager*'s failure to respond within the contract required time limit will lead to deemed acceptance of the *Contractor* submitted Defined Cost.

4.4.3 Fee

In accordance with Clause 52.1, anything that is not covered in the Defined Cost is treated as a Fee. Typically, the types of Fees include Sub-contractor Fee, Direct Fee, and Working Areas Overhead Fee. Section 1 of the Contract Data Part Two defines the percentage of Fee to be multiplied on the Defined Cost, when undertaking the calculation of the Price of Work Done to Date.

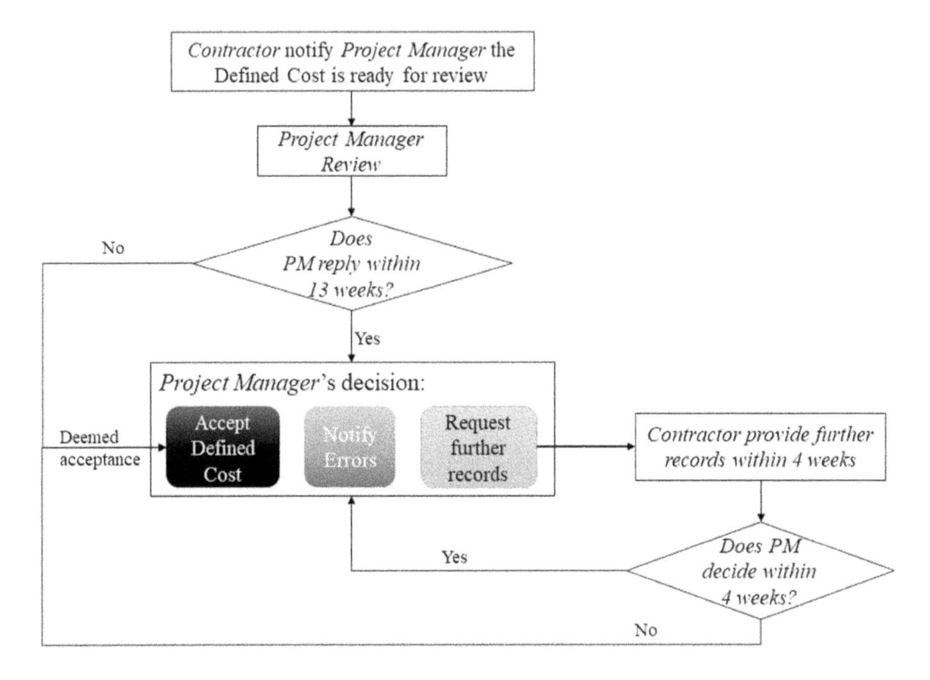

FIGURE 4.10 Defined Cost review process under Clause 50.9 of NEC4 Option C, D, E, and F.

4.4.4 Disallowed Cost

Disallowed cost is the *Contractor* incurred cost, which the *Client* has no obligation to pay. The common example is the People's work outside the Work Areas, and defect correction due to the *Contractor*'s non-compliance or after Completion. As Option A and Option B are fixed price contracts, the Disallowed Cost does not apply to these two options. The Disallowed Cost only applies to Options C, D, E, and F, where the payment depends on the *Contractor*'s actual cost.

Clause C11.2(26) of the NEC4 contract provides the typical definition of the Disallowed Cost as follows:

> *is not justified by the Contractor's accounts and records, should not have been paid to a Subcontractor or supplier in accordance with its contract,*
> *was incurred only because the Contractor did not*
>
> - *follow an acceptance or procurement procedure stated in the Scope,*
> - *give an early warning which the contract required it to give or*
> - *give notification to the Project Manager of the preparation for and conduct of an adjudication or proceedings of a tribunal between the Contractor and a Subcontractor or supplier correcting Defects after Completion,*
>   - *correcting Defects caused by the Contractor not complying with a constraint on how it is to Provide the Works stated in the Scope,*
>   - *Plant and Materials not used to Provide the Works (after allowing for reasonable wastage) unless resulting from a change to the Scope,*

- *resources not used to Provide the Works (after allowing for reasonable availability and utilisation) or not taken away from the Working Areas when the Project Manager requested and*
- *preparation for and conduct of an adjudication, payments to a member of the Dispute Avoidance Board or proceedings of the tribunal between the Parties.*

The payment due will be adjusted by the Disallowed Cost from the Price of Work Done to Date, which applies to the interim payment as well as the final payment.

4.4.5 Retention

If the secondary Option X16 is selected in the contract, the *Client* can retain a percentage of the Price for Work Done to Date (PWDD) in each payment interval or a fixed Fee. In accordance with Clause X16.2, 50% of the retention will be released at Completion or when the *Client* takes over the works and the remaining 50% will be released after the *Supervisor* issues the Defects Certificate.

The fixed retention Fee or the percentage of retention rate is defined in Section X16 of the Contract Data Part One. Figure 4.11 explains the scenario where a fixed percentage of PWDD has been applied as the retention. The typical retention rate is around 3% to 5%, but it can vary depending on the specific provisions of the contract.

4.4.6 Incentivization

The NEC contract also provides several incentive options in order to promote good project management throughout the project life cycle, such as Pain/Gain share under Options C and D.

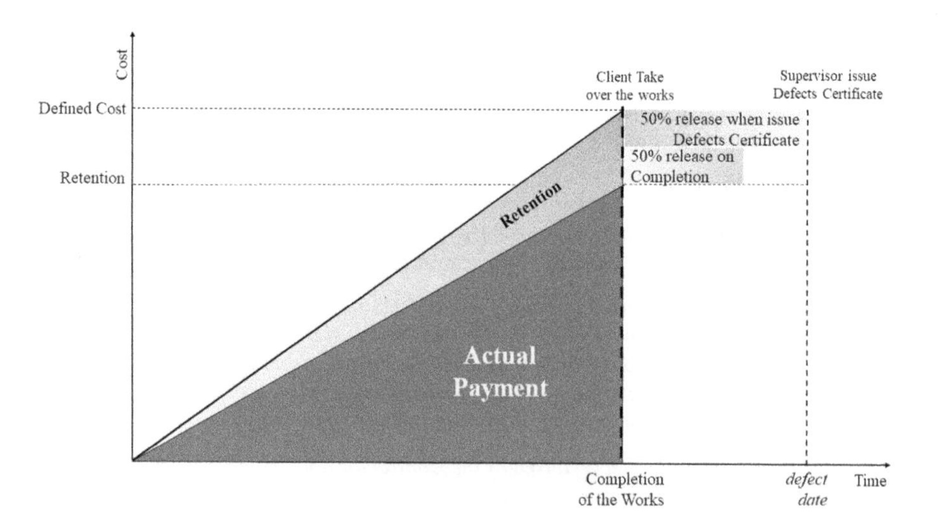

FIGURE 4.11 NEC Retention under Option X16.

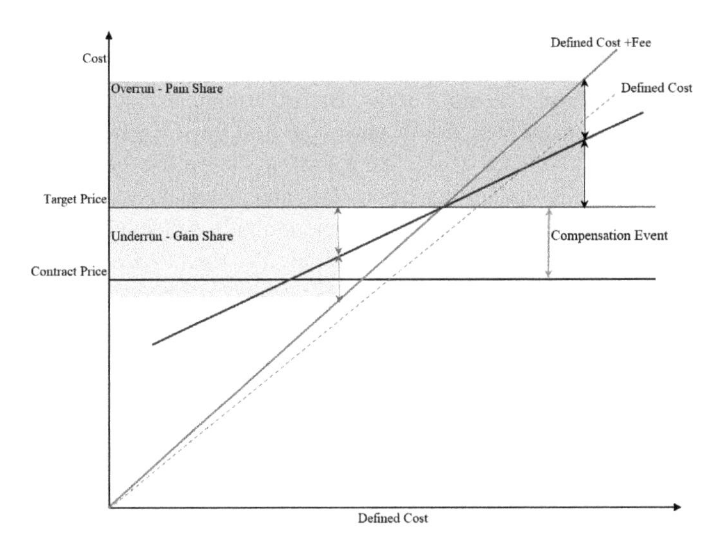

FIGURE 4.12 Pain/Gain Share Mechanism under NEC contract.

4.4.6.1 Target Cost Pain/Gain Share under Options C and D

Figure 4.12 explains the Pain/Gain mechanism under NEC Options C and D. The target cost will be the total of the original contract price plus agreed compensation events. The actual cost is equal to the total Defined Cost plus Fee. The difference between the target cost and the actual cost will then be shared between the *Contractor* and the *Client* in accordance with the ratio defined in the Contract Data.

In addition to the Pain/Gain share, NEC also provides other incentives under the secondary options, e.g., Option X6: Bonus for Early Completion and Option X20: KPI. Option X6 has already been discussed in Chapter 3.

4.4.6.2 Option X20 – Key Performance Indicators (KPIs)

In order to control the *Contractor*'s performance during the project implementation, the *Client* can set out an *Incentive Schedule* in the contract, which requires the *Contractor* to report against the Key Performance Indicators in a reporting interval stated in Section X20 of the Contract Data Part One. If the *Contractor* achieves the target KPIs, then he is entitled to a payment as set out in the *Incentive Schedule.* However, if any of the *Contractor*'s final forecast measurements do not meet the KPI targets required in the *Incentive Schedule*, the *Contractor* needs to submit a proposal to improve his performance to the *Project Manager.* The *Client* may add additional KPIs and associated payment, but cannot delete the existing KPIs or reduce the payment as defined in the *Incentive Schedule* in accordance with Clause X20.5.

4.4.7 Price Adjustment for Inflation

The secondary Option X1 provides the price adjustment for inflation. Due to the long duration of large construction projects, inflation can have a major impact on the *Contractor*'s cost imbursement and profitability. Option X1.1 provides that the payment assessment can be adjusted by the Price Adjustment Factor (PAF). Therefore, the *Contractor*'s cost can be paid with monthly inflation based on the PAF, which is calculated by the base date index at the contract commencement and the index on the payment assessment month. For instance, the Building Cost Information Service (BCIS) of RICS is often used as the benchmark to determine the PAFs. The BCIS index provides monthly updates for labor, plan, and material and the sub-contractor can be used as the benchmark to calculate the monthly adjustment factor in accordance with the proportion of each cost categories. The total PAF adjustment for each month will be the sum of each Defined Cost for each cost category multiplied by the relevant inflation rate of the category.

4.5 COST MANAGEMENT UNDER JCT CONTRACT

JCT 2016 forms of contract provides a broad range of choice for the Employer to select the most appropriate contract type. Table 4.1 classifies the JCT 2016 contract forms under each contract price category.

Large construction projects usually use JCT DB as the form of contract. Unlike the NEC ECC contract, the parties can choose from six main options in accordance with the way the payment is made; the JCT Design and Build 2016 contract is a lump sum contract. Although JCT also provides a wide range of forms of contract, JCT BD is the most used type of contract for large construction projects compared to other forms. In order to overcome the potential rigid character of the fixed price contract, Clause 4.2 of the JCT contract provides various items that may adjust the contract sum as follows:

1. *any amount agreed by the Employer and the Contractor in respect of Changes and other work of the types referred to in clause 5.2 and the amount of each Valuation;*

TABLE 4.1
Contract Price Category of JCT 2016 Forms of Contract

Contract Forms	Abbreviation
Lump Sum Contract	
Standard Building Contract with Quantities	SBC/Q
Standard Building Contract without Quantities	SBC/XQ
Design & Build Contracts	DB
Intermediate Building Contracts	IC
Minor Works Contracts	MW
Major Project Contracts	MP
Remeasurement Contract	
Standard Building Contract with Approximate Quantities	SBC/AQ
Measured Term Contracts	MTC
Prime Cost Contract	
Prime Cost Building Contracts	PCC
Management Building Contracts	MC

2. *any amount agreed by Confirmed Acceptance of an Acceleration Quotation;*
3. *(where the Contract Particulars state that a Fluctuations Provision applies) any amounts payable or allowable under that provision;*
4. *any other amounts referred to in clause 4.12.2 or 4.13.2 (excluding any loss and/or expense to the extent included under clause 4.2.2) and any other deductions referred to in clause 4.12.3 or 4.13.3;*
5. *the deduction of all Provisional Sums included in the Employer's Requirements; and*
6. *any other amount which under this Contract is to be added to the Contract Sum or may be deducted from it.*

4.5.1 Payment

As other contract forms, the JCT contract also provides the rights for the Contractor to be paid by the Employer on time.

4.5.1.1 Interim Payment

Because the JCT contract is primarily used in projects in the UK, the payment provisions of JCT DB 2016 set out in Section 4 "Payment" comply with the Housing Grants, Construction and Regeneration Act (HGCRA) 1996 as amended by the Local Democracy, Economic Development and Construction Act (LDEDCA) 2009. Figure 4.13 demonstrates the interim payment process under JCT forms of contract.

Under Clause 4.9.3, if the payment notice is not issued in accordance with Clause 4.9.2, the amount due will be the sum stated in the interim application for payment.

The first interim Valuation Dates are set out in the Contract Particulars and the remaining interim Valuation Dates are the same date in each month or the closest working day. In the absence of defining the first Valuation Date, it will be one month from when the Contractor possessed the site. The interim Validation Date operates down the supply chain in order to promote the concept of fair payment.

Under Clause 4.7.3, the Contractor needs to submit their interim application for payment based on the "*sum that the Contractor considers to be due to him and the basis on which that sum has been calculated.*"

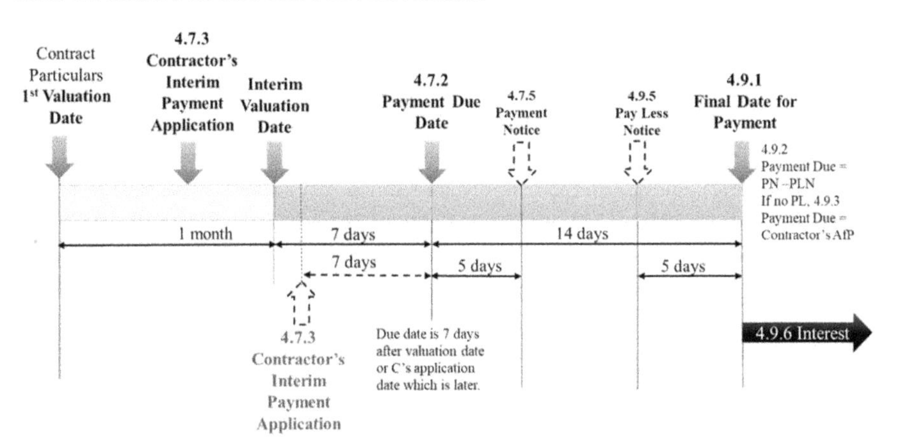

FIGURE 4.13 Interim payment procedure under JCT DB 2016.

The monthly interim due date is seven days after the interim Valuation Date or the date of the Employer receiving the Contractor's payment for application, whichever is later. The due date of the final payment is one month after:

- the end of the Rectification Period for the Works
- the date stated in the Notice of Completion of Making Good
- the date the Final Statement is submitted or the Employer's Final Statement

Within five days of the due date, the *Client* should give a Payment Notice, which specifies the due sum in accordance with Section 110A of the Construction Act 2009 of the Employer's assessment.

In accordance with Clause 4.9.5, if the Employer intends to pay less than the sum stated in the Payment Notice of the interim payment, or the paying party intends to pay less than the sum stated in the Final Payment Notice, the payer should serve a Pay Less Notice (PLN) no less than five days before the final payment due date, which is 14 days after the due date under Clause 4.9.1. As per Sections 111(3) and 111(4) of the Construction Act 2009, the Pay Less Notice should provide the details sum together with the basis of calculation under Clause 4.10.1.

Under Clause 4.9.2, if the Employer provides the valid Payment Notice, subject to the Pay Less Notice, then the payment due is equal to the Payment Notice. However, if the Employer fails to serve the Payment Notice, then the interim payment due will be equal to the Contractor's payment application in accordance with Clause 4.9.3. Likewise, the final payment due sum is equal to the Final Payment Notice in accordance with Clause 4.9.4.

Under Clause 4.7.1, the Employer is obliged to make an interim payment to the Contractor each month. In the case of failure to make payment by the relevant final payment date, the Contractor will be entitled to the interest under Clause 4.9.6 and Clause 4.9.7.

Clause 4.10.3 further provides that the Payment Notice, the Final Payment Notice, the Pay Less Notice, and the Final Statement are required even if the sum due is zero.

4.5.1.2 Final Payment

Upon Completion, the Contractor needs to submit the Final Statement within three months of practical completion in accordance with Clause 4.24.3. If the Contractor fails to submit the Final Statement, the Employer will notify the Contractor that they may issue the Final Statement within two months, unless the Contractor submits their Final Statement before the Employer's submission under Clause 4.24.4. Clause 4.24.2 requires that the Final Statement includes the Contract Sum and the Sum of payment to date. Under Clause 4.24.5, if either Party is dissatisfied with the other Party's Final Statement, they must submit a notice prior to the due date, which is the date one month after either the end of the Rectification Period, the Making Good date in the Notice of Completion, or the date of submission in the Final Statement in accordance with Clause 4.24.5. The Employer shall then give the Final Payment Notice within five days of the due date. The final date for payment is 14 days after the due date in accordance with Clause 4.9.1. In accordance with Clause 4.9.4, the amount due may be subject to the Employer's Pay Less Notice, which must be submitted no less than five days before the final payment date as required by Clause 4.9.5.2. The Employer should make payment in accordance with the Final Payment Notice or Pay Less Notice, whenever it is applicable.

4.5.2 Relevant Matters

JCT deals with the claim for time and money separately. Similar to the Relevant Event to claim the time, JCT sets out five Relevant Matters under Clause 4.21 to allow the Contractor to claim additional money. The Relevant Matters include the following:

1) Change of the contract conditions initiated from the Employer
2) Employer's instruction
3) Archaeological objects
4) Delay by the authorities for necessary permission or approval
5) Any prevention or hinderance by the Employer or its representative

Similar to the procedure to make a time claim, Clause 4.20 sets out the procedure for the Contractor to make a money claim. The Contractor must notify the Employer of a Relevant Matter if it becomes apparent that the Works will lead to loss and/or expenses. The Contractor also needs to notify the Employer of their initial assessment of the loss and/or expenses already incurred and forecast loss and/or expenses of the Relevant Matter in accordance with Clause 4.20.2. The Contractor is required to provide monthly updates of their assessment until the information is sufficient for the Employer to award the total amount under Clause 4.20.3. The Employer must notify the Contractor of the awarded amount of the loss and/or expenses incurred within 28 days of receiving the Contractor's initial assessment or 14 days of each subsequent update in accordance with Clause 4.20.4.

Both the JCT DB 2016 contract and the guidance note for the contract do not make it clear whether the obligation for the Contractor to promptly notify their loss and/or expenses and to provide monthly updates is a condition precedent to its entitlement to the claim of the loss and/or expenses.

In *Bremer Handelgesellschaft v Vanden Avenne* [1978] 2 Lloyd's Rep 109, the House of Lords in the UK set out that for a notice to be a condition precedent it should satisfy the two requirements as follows:

1) state the precise time limit that the notice must be provided
2) make clear that the claiming party will lose its right to claim

In *Walter Lilly & Company Ltd v Mackay and another* [2012] EWHC 1773 (TCC), Clause 4.23.1 of JCT SBC/WQ 1998 provides the requirements for the contract administrator to award the amount of loss and expenses as follows:

> ***provided always*** *that the Contractor shall...make his application as soon as it has become, or should reasonably have become, apparent to him that the regular progress has been or is likely to be affected the contract is which concerned the notification*

> *requirement in relation to loss and expense claims in clause 26.1 of the Standard Form of Building Contract Private Without Quantities 1998.*

Considering the word "*provided always*" illustrates the intention of the Contractor to provide a timely application, therefore the court held that the Contractor's notice is a condition precedent to claim the loss and/or expenses.

Although under JCT DB 2016, the relevant Clause 4.21.1 does include the words "provided always," the Contractor is, however, required to serve the notice as follows:

> *...as soon as the likely effect of a Relevant Matter on regular progress or the likely nature and extent of any loss and/or expense arising from a deferment of possession becomes (or should have become) reasonably apparent to him.*

Clause 4.19.1 provides that the Contractor will be entitled to be reimbursed of loss and expense "*subject to clause 4.19.2 and compliance with the [notification] provisions of clause 4.20.*"

It appears the contract intends to request the Contractor to serve the notice promptly, hence it can be considered as a condition precedent. However, it is uncertain until the decision of the court concludes the consideration.

4.5.3 Claim for Loss and Expense

The JCT contract specifically allows the Contractor to claim for direct loss and/or expenses in accordance with Clause 4.19 under the following circumstances:

- Employer delay of giving the Contractor possession of site under Clause 2.4
- Relevant Matter under Clause 4.21

The loss and/or expenses may include increased prelims, increased overhead, loss of profit, loss of productivity, and interest of financial charge. The Contractor needs to provide the detailed calculation for their estimation. The head office overhead (HOO) may be assessed using the "Hudson," "Emden", or "Eichleay" formulas, which will be discussed in Section 4.6.2.

4.5.4 Fluctuations

For large construction projects with lengthy implementation periods, the Contractor is exposed to increasing risk associated with change of market price and statutory rate. In order to relieve the Contractor of the risks associated with the market or legislation, e.g., inflation, tax, Schedule 7 of the JCT DB 2016 provides provisions for Fluctuations.

The sum of fluctuations is normally calculated based on nationally published price indices, for example CPI, RPI, or industry inflation rate published by the Office of National Statistics (ONS), the Joint Contracts Tribunal (JCT) bulletins, or the Building Cost Information Service (BCIS) index from the Royal Institution of Chartered Surveyors (RICS). Generally, the Contractor cannot claim fluctuations after the Completion Date.

The Fluctuations provision under JCT DB 2016 is the Option A formula, which provides full recovery for all fluctuations.

A Base Date is set out in the Contract Particulars, and the Fluctuation amount is calculated based on the actual works completed to date and applying the relevant indexes to the people, the material and equipment, and the sub-contractor. Meanwhile, the forecast fluctuation is calculated based on the project programme resource profile and cash flow profile with the projected indexes.

4.6 KEY ISSUES

There are some issues associated with the cost management that often become the critical focus of the parties, in particular, the *quantum meruit* which gives the Contractor the rights to get paid for the reasonable sum, which is usually on a reimbursement basis. Furthermore, issues also arise associated with the home office overhead during the period of delay. In addition, the knock-on effect of the recent judgment of the "Smash and Grab" adjudication to the payment of the Contractor has also attracted more attention. The following section explains the legal principles for these burning issues and elaborates the current position of the relevant law.

4.6.1 Quantum Meruit

"Quantum meruit" is a Latin phrase, meaning "*the amount he deserves*." Similar to time at large in the time claim, in construction projects, *quantum meruit* denotes a claim for a reasonable sum for the works and/or service completed and material and equipment supplied. There are four common circumstances that the Contractor can claim payment under *quantum meruit.*

Firstly, if neither the express term nor the implied term in the contract provides a fixed price, the Contractor will be paid under *quantum meruit.* For example, in *Upton-on-Severn Rural District Council v Powell* [1942] 1 All ER 220 (CA), a fireman provided services without an agreed fixed salary, the court awarded reasonable sum to the fireman. *quantum meruit* is also often used in the works undertaken under a letter of intent before the contract incorporation. Furthermore, Section 15 of the Supply of Goods and Service Act (SGSA) 1982 and Section 8 of the Sale of Goods Act (SGA) 1979 both impose the implied terms for the Contractor's entitlement to get paid for a reasonable sum.

Secondly, in the circumstance that the original contract provides a fixed price, but in practice, the scope of work falls outside the original scope of the contract, the Contractor is then entitled to claim payment under *quantum meruit.* For example, in *Steven v Bromley & Son* [1919] 2 KB 722, Steven and Bromley & Son entered into a steel work contract with a fixed rate. However, additional goods outside the original contract were required by Bromley & Son. Steven successfully claimed additional payment for the additional work under *quantum meruit.*

Thirdly, if there is no contract between the parties, for instance the Contractor undertakes the works based on the letter of intent, then the Contractor is entitled to claim payment under *quantum meruit.* For example, in *ERDC Group Ltd v Brunel University* EWHC 687 (TCC), ERDC commenced the works based on a letter of intent and Brunel University was liable for paying ERDC on the basis of *quantum meruit*.

Finally, if the Employer prevents the Contractor's further performance of the work, the Contractor will be entitled to claim payment under *quantum meruit*. For example, in the classic case of *De Barnaby v Harding 1853*, the decedent cancelled the agreement with the claimant without justification. The court held that the claimant can claim the value of service it undertook.

In addition, if the contract parties agreed that the Contractor will be paid with a reasonable sum, the Contractor can also claim on the basis of *quantum meruit*.

Having obtained the entitlement to claim under *quantum meruit*, the next question is what the reasonable sum is. In *Weldon Plant v The Commissioner for The New Towns* [2001] 1 All ER (COMM) 264 (QBD TCC), the court ruled that a fair and reasonable payment should be based on reasonably, properly, and necessarily incurred costs when undertaking the work. Furthermore, in *Serck Controls Ltd v. Drake & Scull Engineering Ltd* [2000] 73 Con. L.R. 100, the court also noted that if the Contractor fails to perform in accordance with the expected productivity and achieve satisfactory quality, then an adjustment to the payment will be made.

4.6.2 Home Office Overhead (HOO)

When claiming costs associated with delays, the indirect cost of home office overhead should also be considered. The most well-used formulas to calculate the home office overhead are the Hudson formula, the Emden formula, and the Eichleay formula.

4.6.2.1 Hudson Formula

The Hudson formula was introduced in the Hudson's Building and Engineering Contracts. Equation 4.1 describes how the home office overhead is calculated by the percentage of the home office overhead and profit stated in the contract, multiplied by the contract daily rate which is equal to contract sum divided by the contract period, and then multiplied by the period of delay.

$$\text{HOO} = \frac{\text{Home Office Overhead \& profit percentage in the Contract}}{100} \times \frac{\text{Contract}}{\text{Contract Period}} \times \text{Period of Delay}$$

Equation 4.1 Hudson Formula for Head Office Overhead (Dennys and Clay, 2015)

The Hudson formula has been well-recognized in the judgements of many cases. However, due to the long duration of large construction projects, the percentage of the home office overhead and profit stated in the contract may not reflect the real situation in the whole project life cycle.

4.6.2.2 Emden Formula

In order to cope with the potential limitations in the Hudson formula, the Emden formula was published in Emden's Building Contracts and Practice. Similar to the

Hudson formula, the Emden formula kept the majority of the components of the Hudson formula, but replaces the percentage of the home office overhead and profit stated in the contract with the actual percentage of the home office overhead and profit, as described in Equation 4.2.

$$HOO = \frac{\text{Actual Home Office Overhead \& profit Percentage}}{100} \times \frac{\text{Contract Sum}}{\text{Contract Period}} \times \text{Period of Delay}$$

Equation 4.2 Emden Formula for Head Office Overhead (Emden, Bickford-Smith, and Freeth, 1980)

The actual percentage of the home office overhead and profit is calculated by the total home office overhead cost of the Contractor divided by the Contractor's total revenue. Since published, the Emden formula has received wide support in various court cases.

4.6.2.3 Eichleay Formula

Both the Hudson formula and the Emden formula rely on the proof of loss of opportunity by the Contractor. The Eichleay formula was developed to fill the gap between the Hudson formula and the Emden formula. There are typically three steps to calculate the home office overhead using the Eichleay formula.

Step 1 is to calculate the Contractor's allocable overhead, as described in Equation 4.3.

$$\text{Allocable Overhead} = \text{Total HOO for Contract Period} \times \frac{\text{Total Contract Billing}}{\text{Total Company Billing over the Contract Period}}$$

Equation 4.3 Step 1 of the Eichleay Formula for Head Office Overhead (SCL, 2017)

First, the percentage of the existing project in the whole portfolio of the Contractor's company is calculated by dividing the total project cost by the total revenue of the Contractor's company during the period of the contract. Then the Contractor's allocable overhead is calculated with the project percentage in the Contractor's company by multiplying by the total home office overhead during the contract period.

Step 2 then calculates the Daily Overhead Rate for the contract, as described in Equation 4.4.

$$\text{Daily Overhead Rate for Contract} = \frac{\text{Allocable Overhead}}{\text{Contract Period}}$$

Equation 4.4 Step 2 of the Eichleay Formula for Head Office Overhead (SCL, 2017)

The Daily Overhead Rate is calculated by dividing the total contract duration by the Allocable Overhead, which was established in Step 1.

Step 3 finally calculates the unabsorbed overhead with the Daily Overhead Rate calculated in Step 2 by multiplying the period of delay, as described in Equation 4.5.

$$\text{Unabsorbed Overhead} = \text{Daily Overhead Rate} \times \text{Period of Delay}$$

Equation 4.5 Step 3 of the Eichleay Formula for Head Office Overhead (SCL, 2017)

In order to make a clearer comparison of these three formulas and help readers to understand these formulas, the consolidated calculation of the Eichleay Formula is illustrated in Equation 4.6.

$$\text{HOO} = \frac{\text{Total HOO for Contract Period}}{\text{Total Company Billing over the Contract Period}} \times \frac{\text{Contract Sum}}{\text{Contract Period}} \times \text{Period of Delay}$$

Equation 4.6 Consolidated Eichleay Formula for Head Office Overhead

It is now clear that the difference between the Hudson formula, the Emden formula, and the Eichleay formula is the first part of calculating the rate of the home office overhead. The parties can then choose the appropriate rate for the project undertaken when calculates the home office overhead for the period of delay. The parties can then choose the relevant equation accordingly.

4.6.3 "Smash and Grab"

HGCRA 1996 and the following amended Construction Act 2009 requests the Employer to pay the Contractor on time and argue later. In accordance with the Construction Act 2009, the first payment due date is five weeks after the contract start. If the Employer believes the Contractor applied for more than they deserve, then the Employer should serve a Pay Less Notice (PLN) after the payment due date. This Pay Less Notice should be served no less than seven days before the payment date. If the Employer fails to serve the payment notice, then the Contractor can send a default payment notice and the Employer has to pay in accordance with the payment notice. This approach echoes the "pay now and argue later" philosophy of the Construction Act 2009.

The phrase "Smash and Grab" originated from *CG Group Limited v Breyer Group PLC* [2013] EWHC 2959 (TCC). It is a claim within the circumstances of the Employer or their representative failing to serve the payment notice or Pay Less Notice, which gives the right to the Contractor to be paid in accordance with their application for payment. For example, in *SG Construction Ltd v Seevic College* [2014] EWHC 4007 (TCC), the court held that:

> *...if the employer fails to serve any notices in time it must be taken to be agreeing the value stated in the [interim] application, right or wrong...*

Therefore, if the Employer fails to serve the payment notice within the time limited as required under the Construction Act 2009, the Contractor is entitled to the payment as their application for payment. Likewise, in *Galliford Try Building Ltd v Estura Ltd* [2015] EWHC 412 (TCC), the court followed the approach in *ISG* and ruled that:

> *if an employer [the payer] fails to serve the relevant notices...it must be deemed to have agreed the valuation stated in the relevant interim application, right or wrong.*

However, this "Smash and Grab" approach was not followed for the payment associated with the final account. In *Matthew Harding (t/a MJ Harding Contractors) v Paice* [2015] EWCA Civ 1231, the court ruled that the Employer can commence a second adjudication to determine the true value of the final account. This decision is further supported by Mr Justice Coulson in *Grove Developments Ltd v S&T (UK) Ltd* [2018] EWHC 123 (TCC). In this case, S&T submitted its application for an interim payment of £14m; Grove then responded with a payment certification of £1.4m together with the evaluation sheet. However, because Grove issued the Pay Less Notice three days later than the time required under the Construction Act 2009, S&T contended that the Pay Less Notice was invalid. Consequently, S&T was entitled to the payment of £14m. Under the "Smash and Grab" approach, the Employer can only start the adjudication for the true value of the work, until they have paid £14m. However, the English Court of Appeal held that Grove can initiate a second adjudication for the true value of the interim payment without paying the £14m as requested by the Contractor.

The Court of Appeal's departing from "Smash and Grab" in *Grove v S&T,* saved Grove from a potentially significant financial deficit. On the one hand, it urges the parties to comply with the procedure requirement under the Construction Act 2009; on the other hand, it demonstrates the court's intention to seek justice for the parties.

4.6.4 Inflation Impact

Because large construction projects usually take place over several years, inflation adjustment may have an important impact on the Contractor's profit. For example, the Crossrail project in London took over ten years to complete. Based on the inflation rate published by the Office for National Statistics (ONS) over the past ten years, assuming the costs are accrued evenly, the Contractor may be able to get paid 13.86% inflation of the whole contract value. Considering the typical project spending profile, the range of inflation entitlement for the Contractor in the UK is compounded at about 3% per year.

However, if the project is undertaken under Option C or D of the NEC contract, which will share the Pain/Gain, the Contractor's benefit associated with the inflation adjustment may be undermined if the project is subject to loss. For example, if a large infrastructure project takes ten years and the original target contract price is £10 billion, assuming the cost is spread evenly during the project and the inflation adjustment rate is 3% per year, even if the Contractor's actual cost is the same as the original contract price, due to the impact of compound inflation, the Contractor will be paid £1.46 billion more, which is 14.6% more than the original contract Price, as illustrated in Table 4.2.

TABLE 4.2
Sample Inflation Impact to Long-Term Project

	Year1	Year2	Year3	Year4	Year5	Year6	Year7	Year8	Year9	Year10	Total
Budget (£000,000)	1,000	1,000	1,000	1,000	1,000	1,000	1,000	1,000	1,000	1,000	10,000
Actual Cost (backdated) (£000,000)	1,000	1,000	1,000	1,000	1,000	1,000	1,000	1,000	1,000	1,000	10,000
Years of Compound	0	1	2	3	4	5	6	7	8	9	
Inflation %	3.0%	3.0%	3.0%	3.0%	3.0%	3.0%	3.0%	3.0%	3.0%	3.0%	
Inflation adjustment factor	0%	3%	6%	9%	13%	16%	19%	23%	27%	30%	
Inflation (£000,000)	0	30	61	93	126	159	194	230	267	305	1,464
Total Actual Cost (£000,000)	1,000	1,030	1,061	1,093	1,126	1,159	1,194	1,230	1,267	1,305	11,464

However, when calculating the Pain/Gain, the cost management team often ignores the potential inflation impact to the original contract target price. In this case, the overpaid inflation adjustment of £1.46 billion will be considered as a loss, which will be shared between the *Contractor* and the *Client* under Option C or Option D of the NEC contract. Assuming the Pain/Gain share ratio between the *Client* and the *Contractor* is 50:50, then the *Contractor* will end up with a share of £0.78bn, which they are actually entitled to.

In order to avoid such an injustice, the *Contractor*'s actual cost should be adjusted and backdated to the start of the project. Alternatively, the inflation adjustment should be added on top of the target price together with the approved compensation event when calculating the Pain/Gain.

4.7 RECOMMENDATIONS

Project overrun is a common issue for projects across all industries. In general, only about 70% of projects are able to complete within the budget. For large construction projects, the probability is even lower, particularly for rail projects: only about 50% of projects finish within the budget. Therefore, it is important to control the cost throughout the project. The financial system used to manage the project account and cost can have an important effect on the effectiveness of project cost control. It is important the set up appropriate procedures at the beginning of the project and ensure the cost management staff comply with the procedure.

In order to control the project actual cost within the project budget, it is important to set up the project cost breakdown structure in line with the work breakdown structure defined in the Scope of work as well as in the Programme. In practice, due to the complexity of large construction projects, it is difficult to record cost information as detailed as work breakdown structure. In such circumstances, it would also beneficial to set the cost breakdown structure in line with the higher level of the work breakdown structure.

In the event of a Joint Venture or Consortium, due to the actual financial or commercial management system used by the Contractor/s, it may be difficult to arrange the project cost breakdown structure to suit the work breakdown structure. The alternative way to establish good control of the project is to define the cost account to each resource allocated to the project activities. The earned value can then be calculated at the resource level and exported to Excel from the planning software, e.g., using Primavera P6 to re-categorize it into the cost account in line with the financial system. The earned value and actual cost can then be compared appropriately under each cost account.

Due the impact of the statutory payment time limit, the payment process under the NEC contract is around one month, which is much shorter than the payment cycle under FIDIC. In addition, the Contractor gets paid for about one to two months forecast Defined Cost under Option C, D, and E. If the Contractor's financial position is not strong enough and requires better cash flow to support the project implementation, then it is better to choose the NEC contract.

Due to the long duration of large construction projects, staff turnover is common throughout the project life cycle. It is important to establish a centralized financial system to capture all records associated with any costs occurred in the project. If the main contractor constitutes a Joint Venture, the situation may be more complicated. In such circumstances, a cost management system and timesheet booking system should be established for the project as soon as possible after the contract is awarded. It is also worth undertaking annual internal and/or external audits to identify any potential issues and take corrective action in time if necessary.

The cost management team should track the actual cost in time, record the cost incurred in the correct cost account, and take control of the costs against the original budget. If there is any variance associated with the scope or the quantities, it should be recorded clearly and communicated with the project manager or cost manger. The cost management team should also work closely with the change control team and planning team, in order to control the project more effectively.

Inflation is important for the contract to have a payment adjustment for the long-term project; the Contractor should ensure the inflation provisions are included in the contract during drafting stage. However, potential issues associated with the inflation payment (e.g., Target Price with Pain/Gain) can create an injustice between the parties. The contract parties should consider the potential impact in accordance with the relevant contract used and adjust the inflation impact to the relevant calculation in the final account.

Furthermore, the statistics show that waste in the construction industry is a common feature. Even for the country with the least waste, China generates about 7% waste in construction projects. In the UK, the waste rate is about 8% for construction projects. Therefore, if the waste can be controlled effectively, the overall project overrun will be improved and the Contractor can make more profit in the lump sum contract.

REFERENCES

Book/Article

Axelos. 2017. *Managing Successful Projects with PRICE2*. 2017 Edition. Norfolk, UK: TSO (The Stationery Office) / Williams Lea Tag.

Dennys, N. and Clay, R. 2015. *Hudson's Building and Engineering Contracts*. 13th Edition. London, UK: Sweet & Maxwell Ltd.

Emden, A. E., Bickford-Smith, S. and Freeth, E. 1980. *Emden's Building Contracts and Practice*. 8th Edition. London, UK: Butterworths.

PMI. 2017. *A Guide to the Project Management Body of Knowledge, PMBOK Guide*. 6th Edition. Newtown Square, PA: Project Management Institute.

SCL. 2017. *Delay and Disruption Protocol*. 2nd Edition. London, UK: The Society of Construction Law.

Contract

FIDIC. 1999. *Conditions of Contract for Construction*. 1st Edition. (1999 Red Book). Geneva, Switzerland: The Fédération Internationale des Ingénieurs-Conseils.

FIDIC. 2017. *Conditions of Contract for Construction*. 2nd Edition. (2017 Red Book). Geneva, Switzerland: The Fédération Internationale des Ingénieurs-Conseils.

FIDIC. 2017. *Conditions of Contract for EPC / Turnkey Project*. 2nd Edition. (2017 Yellow Book). Geneva, Switzerland: The Fédération Internationale des Ingénieurs-Conseils.

FIDIC. 2017. *Conditions of Contract for Plant & Design Build.* 2nd Edition. (2017 Silver Book). Geneva, Switzerland: The Fédération Internationale des Ingénieurs-Conseils.
JCT. 2016. *Design and Build Contract 2016.* London, UK: The Joint Contracts Tribunal Limited.
JCT. 2016. *Standard Building Contract with Quantities 2016.* London, UK: The Joint Contracts Tribunal Limited.
NEC. 2013. *NEC3 Engineering and Construction Contract.* London, UK: Thomas Telford Ltd.
NEC. 2017. *NEC4 Engineering and Construction Contract.* London, UK: Thomas Telford Ltd.

Statutes

Supply of Goods and Service Act 1982.
Sale of Goods Act 1979.

Cases

CG Group Limited v Breyer Group PLC [2013] EWHC 2959 (TCC).
ERDC Group Ltd v Brunel University EWHC 687 (TCC).
Galliford Try Building Ltd v Estura Ltd [2015] EWHC 412 (TCC).
Grove Developments Ltd v S&T (UK) Ltd [20180 EWHC 123 (TCC).
Matthew Harding (t/a MJ Harding Contractors) v Paice [2015] EWCA Civ 1231.
Serck Controls Ltd v. Drake & Scull Engineering Ltd [2000] 73 Con. L.R. 100.
SG Construction Ltd v Seevic College [2014] EWHC 4007 (TCC).
Sharpe v San Paulo Railway (1873) LR 8 Ch App 597.
Steven v Bromley Son Bankes [1919] 2 KB 722.
Upton-on-Severn Rural District Council v Powell [1942] 1 All ER 220 (CA).
Weldon Plant v The Commissioner for The New Towns t [2001] 1 All ER (COMM) 264 (QBD TCC).

5 Risk Management

5.1 INTRODUCTION

Due to the complexity and long duration of large construction projects, uncertainty is inevitable throughout the project life cycle. Risk management is an important component of project management. This chapter starts by introducing the risk management principles under the Project Management Body of Knowledge (PMBOK) and the Standard for Risk Management in Portfolios, Programs, and Projects of the Project Management Institute (PMI) and Risk principle under PRINCE2. It then discusses how risk allocation has been arranged under the NEC and FIDIC forms of contract respectively and the procedure to undertake risk management under both contracts. The chapter then discusses how the quantitative cost and schedule risk analysis has been undertaken in complex construction projects and how it can be of benefit to the overall project. The chapter concludes with recommendations for best practice of risk management in large construction projects.

5.2 STANDARD RISK MANAGEMENT PROCEDURE

PMI (2019) defines risk is "*an uncertain event or condition that, if it occurs, has a positive or negative effect on one or more objectives.*" According to PMI's 2015 Pulse of the Profession report, when using formal risk management in an organization, the overall project performance improves approximately 15% in terms of achieving the original objectives, finishing on time, and completing within budget (PMI, 2015). Therefore, risk management plays an important role in successful project management. It ensures proactive management of issues and changes, and consequently improves decision making, and increases realization of opportunities.

Risk management is an important component of project management. It plays a significant role in the project's success and the success of risk management starts with a good risk management strategy and plan. Key project management organizations established risk management standards or guidelines to facilitate project management professionals to effectively manage risks. For example, the two well-known project management standards of PMBOK and PRICE2 both provide detailed risk management procedures. PMBOK (PMI, 2017) describes project risk management as a process of "*conducting risk management planning, identification, analysis, response planning, response implementation, and monitoring risk on a project.*" PRINCE2 (Axelos, 2017) provides a five-step risk management procedure, including identify, assess, plan, implement, and communicate. Further, HM Treasury of the United Kingdom also publishes the Orange Book to provide risk management principles and procedures for UK government projects. Table 5.1 describes the standard risk management procedures under PMBOK, PRINCE2, the Orange Book of HM Treasury, and ISO3100.

PMBOK of the Project Management Institute (PMI) recommends that risk management begins by planning risk management. A risk management plan is usually

TABLE 5.1
Standard Risk Management Procedures

PMBOK	PRINCE2	HM Treasury Orange Book	ISO3100
Identify	Identify	Identify	Identify
Analysis	Assess	Assess	Analysis
Response	Plan	Address	Treatment
Monitor & Control	Implement	Review & Report	Monitor & Review Record & Report
Communication	Communication	Communication & Learning	Communication & Consultation

a component of the project management plan or the project execution plan which needs to be developed at the beginning of the project. The on-going risk management procedure then includes risk identification, risk analysis, risk response, risk mitigation, and risk monitoring.

Under PRINCE2 (Axelos, 2017), risk management is undertaken by four sequential steps which are to identify risks, assess the potential impact, plan mitigation, and implement mitigation actions. Finally, the on-going communication as the fifth step interacts with all stakeholders across the four sequential steps above. As the government projects in the UK largely apply PRINCE2 principles, the Orange Book of HM Treasury follows a similar approach to the PRINCE2 risk management process.

In addition, ISO31000 also establishes a specific standard for risk management, which consists of risk management principles, framework, and process. The risk management process under ISO31000 starts with scope, context, and criteria, followed by risk assessment which includes risk identification, risk analysis, and risk evaluation; the risk treatment is then continued afterwards. These are standard processes that will be repeated in several rounds during the project. Meanwhile, the on-going processes to facilitate risk management are carried out throughout the project life cycle, including recording and reporting, monitoring and review, and communication and consultation.

In summary, standard risk management includes the four main processes of risk identification, risk analysis, risk treatment, and risk monitoring and control. Besides, the communication is undertaken by the project stakeholders throughout the project life cycle.

5.2.1 Risk Identification

Risk identification is the foundation of the subsequent risk management, as only the identified risks can be analyzed and managed further. In practice, risk identification largely relies upon expert judgement. Most risks are initially identified through a brainstorming session in a risk workshop at the beginning of the project. The risk manager or the project manager then reviews the outcomes from the risk workshop and prepares a risk register for risk management. The risk manager may also identify risks from other sources, e.g., lessons learned, risk breakdown structure, cause and effect diagram, root-cause analysis, SWOT analysis, checklist, questionnaire, interview, industry knowledge base, and assumption and constraint analysis. The additional identified risks will be recorded in the risk register and then shared with the project team for a decision. If the budget and resource are available, the Delphi technique may also be used for risk identification. Delphi collects the expert opinion through several rounds of questionnaires. The risk manager summaries the experts' response from the previous round and prepares the questionnaire for the subsequent round to seek further clarification and understand the potential risks deeper; the experts make further decisions based on the summarized outcomes from the previous round and finally decide their response during the process. This avoids the potential disadvantage of a dominant opinion during the opening brainstorming session in a risk meeting or workshop.

5.2.2 Risk Analysis

When a risk list is identified and input to the risk register, the next stage is to analyze these risks. Risk analysis includes two steps: first a qualitative analysis will be undertaken for all of the risks captured in the risk register. For the risks of high impact and likelihood, further quantitative analysis may also be undertaken.

5.2.2.1 Qualitative Analysis

Qualitative risk analysis provides a preliminary estimate of the overall effect of risk on the project. The analysis results are then used to screen important risk elements for escalation and determine the priority for each risk in the following management. The most common approach to undertake qualitative risk analysis is risk probability and impact assessment, which evaluate the risk impact and risk likelihood based on the existing evaluation matrix set up in the risk management plan and risk register

TABLE 5.2
Typical Qualitative Risk Evaluation Category

Category	Very Low (VL)	Low (L)	Medium (M)	High (H)	Very High (VH)
Cost	<1%	1% to 2.5%	2.5% to 5%	5% to 10%	>10%
Schedule	<1 day	1–7 days	1–2 weeks	2–4 weeks	>1 month
Performance	>95%	85–95%	75–85%	65–75%	<65%
Health & Safety	Near Miss	Incident	Reportable Injury	Major Injury	Fatality
Quality	Minor Defects	Minor defects on secondary operation	Defect impact on key functionality	Major defect restricts key operation	Major defect prevents key operation

template. The risk impact of each project component can be defined separately, for instance, time, cost, quality, health and safety, and reputation. Table 5.2 demonstrates a sample risk matrix to undertake qualitative risk analysis.

In accordance with the specific features of the project, specific categories may also set up, for instance, quality, reputation, service broken duration, etc. The risk analysis at this stage may still be impacted by the participants involved in this process and the result of analysis is generally in a relative broad grade. The risk evaluation categories shown in Table 5.2 are usually converted to a typical five grid risk analysis matrix, which is used to evaluate the importance of risk elements, as shown in Figure 5.1.

The project manager or risk manager will usually focus on the risks that appear in the top right range in the risk matrix and escalate those to the senior management. Meanwhile, these risks will be treated with high priority and the comprehensive quantitative risk analysis may be taken for these risk elements.

The qualitative risk analysis is usually undertaken for both pre-mitigation and post-mitigation scenarios, as shown in Figure 5.2 The project manager can then

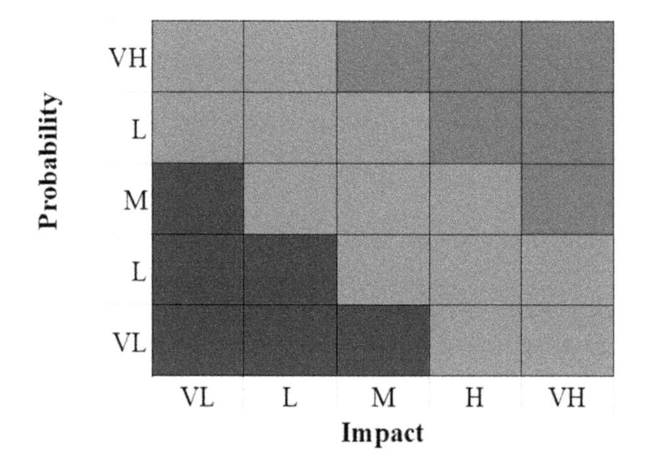

FIGURE 5.1 Qualitative risk analysis – probability and impact risk matrix.

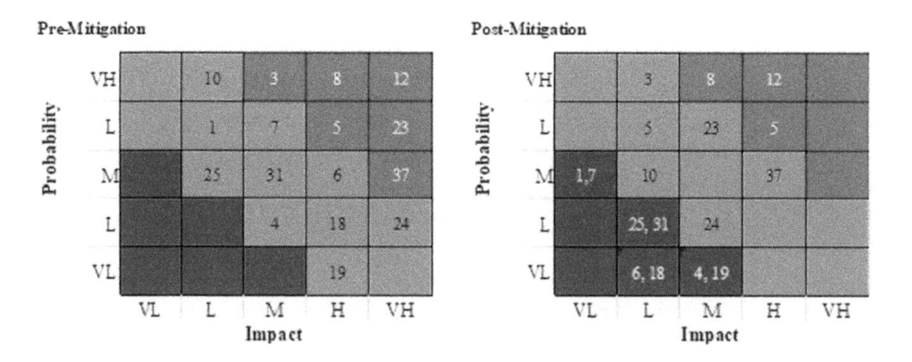

FIGURE 5.2 Compare pre-mitigation and post-mitigation risk profile.

decide the effectiveness of the mitigation actions and any further actions if required. It also provides a dynamic view of the current status of risks and elements on which to focus.

Other qualitative analysis methods may also be used for qualitative analysis, for example affinity diagram, influence diagram, Analytic Hierarchy Process (AHP), and Nominal group technique.

5.2.2.2 Quantitative Analysis

The quantitative risk analysis is usually used for predicting the potential impact based on the combined effects of the risks. It defines the distribution profile for single risk element, uses the schedule or cost model, and applies advanced statistic software to estimate the risk impact under different scenarios.

Based on the complexity and the importance of the risk, some risks may require the need to undertake comprehensive quantitative analysis. The most common quantitative risk analysis method is Monte Carlo simulation. The simulation for large projects can be very complicated, particularly for the schedule risk analysis. Sections 5.6 and 5.7 will discuss quantitative cost risk analysis (QCRA) and quantitative schedule risk analysis (QSRA) for complex projects in further detail.

Following the comprehensive Monte Carlo analysis for schedule risk and cost risk, the primary options can be input into a decision tree to facilitate the senior management to make final go/no-go decisions, as shown in Figure 5.3.

Other less complicated quantitative analysis may also be applicable to less critical risk elements, e.g., decision tree analysis, Latin Hypercube Sampling (LHS), Contingency Reserve Estimation, Expected Monetary Value (EMV), Failure Modes and Effects Analysis (FMEA) or fault tree analysis, and Programme/Project Evaluation and Review Technique (PERT).

Among these methods, Expected Monetary Value (EMV) analysis is often used in the early stage of a project to determine the contingency reserve. EMV is a statistic technique that calculates the expected cost or benefit in accordance with its probabilities of occurring for each potential outcome. The formula for calculating EMV is described in Equation 5.1.

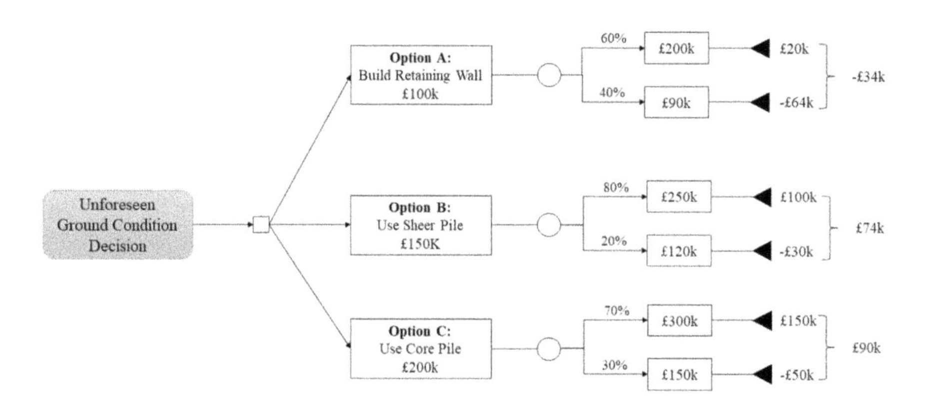

FIGURE 5.3 Decision tree analysis.

$$\text{Expected monetary value}(\text{EMV}) = \sum_{i=1}^{n} \text{Probability}_i \times \text{Impact}_i$$

Equation 5.1 Expected Monetary Value

EMV is based on a weighted average for each scenario by multiplying the probability with cost impact. It then sums up all value of identified outcomes. EMV provides the average risk value for a single event in the combined outcome profile.

5.2.3 Risk Treatment

After analyzing the impact of risks, the next stage is to prepare the risk response strategy and implement mitigation actions. In practice, risk management not only identifies and controls risks, but also manages opportunities. Risk treatment includes planning the response of the identified risks, and implementation of mitigation actions.

Under PMI PMBOK (2017), the response strategy to threats including Avoid, Transfer, Mitigate and Accept, and the response approach for opportunity includes Exploit, Share, Enhance, and Accept.

PRINCE2 (Axelos, 2017) applies PASTAR for threat response and PASTEE for opportunity response. Prepare contingency plans, Accept, Share, and Transfer are applicable to both threat response and opportunity response. Then, Avoid and Reduce are specifically used for threat response, whereas Exploit and Enhance are applied to opportunity response.

Likewise, the risk treatment options under ISO31000 include avoiding risk, removing causation sources, changing the probability and/or the consequences, sharing through contracts or insurance, and simple retain (ISO, 2018).

If the Oracle Primavera Risk Analysis (PRA) is used for the overall risk management, the PRA integrated risk register provides four categories of risk responses, including: Avoid, Transfer, Reduce, and Accept.

Once the risk treatment strategy is developed, then competent personnel will be delegated as the action owners, who will undertake the relevant actions by the required time limit to mitigate the risks or enhance the opportunities. It is good practice to also allocate accountable personnel to ensure the agreed actions are proceeded by the action owner or an alternative delegated competent person.

5.2.4 Risk Monitoring and Control

Following the implementation of risk treatment strategy, the risk manager then needs to monitor and control the progress of the mitigation actions and update with the residual risks or secondary risks following the risk mitigation. In these circumstances, an additional risks treatment approach may need to be decided for new risks. Furthermore, the overall risk status and top risks will be reported in the period project progress report and discussed with the senior management team and key stakeholders.

5.3 RISK MANAGEMENT UNDER NEC FORMS OF CONTRACT

NEC encourages proactive risk management. NEC intends to allocate risk to the Party who can best manage it. Meanwhile, the Parties successfully undertaking of risk management will result in financial benefit.

5.3.1 Risk Allocation

The NEC contract allocates risks through Core Clause Section 8 "Liabilities and Insurance" and Section 6 "Compensation Event," main options, secondary options, and the Contract Data.

5.3.1.1 Main Options

When the Parties enter into a contract, the first question to consider is the main options of the contract. The *Contractor*'s risk increases with increasing of the alphabetical order of the main options, whereas the *Client*'s risk decreases. Therefore, the *Contractor* bears the highest risk under Option A and bears the lowest risk under Option E. In contrast, the *Client* bears the highest risk under Option E and bears the lowest risk under Option A. Under Option C, the Parties only share the risks on cost under the Pain/Gain share, but do not share the risks on time.

The contract provision related to risk allocation between the Parties includes Contract Data, Core Clause 8, and compensation event under Core Clause 6.

5.3.1.2 Core Clause

Core Clause 80.1 defines the *Client*'s liability under NEC4, which is changed from *Employer*'s risks under NEC3. It provides the list of categories for which the *Client* bears risks, which include:

- use or occupant of the Site
- fault of the *Client* or the *Client*'s design
- loss of or damage to Plant and Materials supplied by the *Client* before the *Contractor*'s acceptance
- loss of or damage to the Work, Plant and Material due to *force majeure*
- loss of or damage to the Works taken over by the *Client*
- loss of or damage to the Equipment, Plant and Material after termination
- loss of or damage to the property owned by the *Client*
- additional liability specified in the Contract Data part two section 8

Under NEC3 (NEC, 2013), Clause 81.1 provides that "*the risks which are not carried by the Employer are carried by the Contractor.*" This means, apart from the risks expressly provided under Clause 81.1, Contract Data part one, and compensation events under Clause 60.1, all other risks are carried by the *Contractor*. NEC4 revised this clause and listed the specified *Contractor*'s risks including:

- claims and proceedings from Others
- loss of or damage to the property owned by the *Client*
- loss of or damage to the property owned by the *Client*, but related to the Works provided by the *Contractor*
- death or injury of the *Contractor*'s employee

The change of Clause 81.1 certainly relieves the *Contractor* from enormous uncertainty of the unidentified risks. It is fairer to the contract Parties and it maintains the contract's spirit of mutual trust and cooperation.

5.3.1.3 Contract Data

Under the last item of Section 1 "General" of the Contract Data part one and part two under the NEC4 contract, or the sixth bullet point of "statements given in all contracts" of the Contract Data part one of NEC3, the *Contractor* should provide the matters that will be included in the Early Warning Register.

Section 8 "Liabilities and Insurance" of the Contract Data part two specifically lists the additional *Client*'s risk for the current project.

5.3.1.4 Compensation Event

In addition, the Core Clauses under 60.1 of compensation event further provide 21 components for which the *Client* bears the risks, as discussed further in Chapter 6.

Clause 60.1(14) gives the rights to the *Contractor* to claim compensation for the *Client*'s risks defined in Clause 80.1 and the Contract Data part one section 8.

The unforeseen ground condition is one of the most significant risks in large construction projects. In accordance with Clause 60.1(12), the *Contractor* is entitled to claim compensation event for the physical conditions above the defined risk. Therefore, the *Client* bears the risk for unforeseen ground conditions. It is worth emphasizing that the *Contractor* should include the risk allowance for both time and cost in the quotation of the compensation event.

5.3.1.5 Secondary Options

Contract Parties can also use the secondary options to allocate certain risks. Table 5.3 describes the secondary options that can be used to allocate the *Client*'s risks and/or the *Contractor*'s risks.

TABLE 5.3
Risk Allocation Under Secondary Options of NEC Contract

Ref	Option	Risk allocated
The *Client*'s Risks		
X1	Price adjustment for inflation	Cost
X2	Changes in the law	Legal
X3	Multiple currencies	Cost
X6	Bonus for early Completion	Cost & Time
X14	Advanced payment to the *Contractor*	Cost & Legal
X15	The *Contractor*'s design	Cost & Legal
X18	Limitation of liability	Cost & Legal
The *Contractor*'s Risks		
X4	Ultimate holding company guarantee	Legal
X5	Sectional Completion	Cost & Time
X7	Delay damages	Cost & Time
X13	Performance Bond	Cost
X16	Retention	Cost
X17	Low performance damages	Cost
Shared Risk between the *Client* and the *Contractor*		
X20	Key Performance Indicators	Cost & Legal

On the one hand, the *Contractor* can transfer the inflation risk to the *Client* by including Option X1 "Price Adjustment for Inflation," or the risk of law changes by selecting Option X2 "Changes in the Law." If Option X15 "the *Contractor*'s Design" is selected, the *Contractor*'s design liability will be restricted to the reasonable skill and care instead of "fitness to purpose" under most design and build contracts. Furthermore, Option X18 "Limitation of liability" sets out the *Contractor*'s liability to the *Client* in terms of indirect or consequential loss, loss of or damages to the *Client*'s property, Defects due to the *Contractor*'s design outside the Defects Certificates, the total liability, and other matters notified to the *Contractor* before the liability period expires.

On the other hand, the *Client* can use Option X4 "Ultimate holding company guarantee" to mitigate the risk to the *Contractor*'s insolvency. The *Client* can also use X5 "Sectional Completion" to set up phased completion for different components of the project in order to take over certain works early or start operation certain services early. Sectional completion can also set up works in stages to incentivize the *Contractor* to deliver on time, for example Civils works, M&E work, and Fit-out, and Test and Commissioning. This will force the *Contractor* to complete each phase on time and thus not delay the subsequent phase. Option X7 "Delay Damages" and Option X17 "Low Performance Damages" are often used to sanction the *Contractor*'s failure to perform as required by the Completion Date, or performance rate under the contract. This allows the *Client* to obtain potential losses due to the delayed use of the end product.

Option X13 "Performance Bond" provides the *Client* the compensation in the event of the *Contractor*'s failure to perform or insolvency. There are two type of performance bond: bank bond and insurance bond. The *Client* can call the bank bond at any time and the bank would not undertake any queries or investigation when the *Client* calls for the bond. This type of bond is the most favorable to the *Client*. However, the bank bond usually has a high application thread hold, and thus only limited tier one *Contractors* can obtain such a bond. Therefore, the insurance bond is often used for construction projects. However, the insurance company will undertake an investigation when the *Client* calls for the bond. Therefore, the *Contractor* is exposed to less risk to the potential of the bond being called off.

Option X16 "Retention" is another way to protect the *Client* from the risk of any uncorrected defect, disallowed cost to be justified, or uncompleted works, etc. The retention is usually deducted as a fixed percentage of the *Contractor*'s monthly payment application. Under Clause X16.2, upon completion of the whole works, 50% of the retention is released and the other half will be released when issuing the Defects Certificate, usually after the *defect date*.

Option X20 "Key Performance Indicators" is often used in large construction projects to incentivize the Parties' performance. A series of Key Performance Indicators (KPI) has been set up in the contract and reported on a monthly basis usually in line with the payment intervals. The incentive values are then calculated based on the KPI reported in each period.

In addition, the contract Party can also agree any additional risks that gives the *Contractor* the ability to seek compensation under Z clause. However, it is highly recommended that the contract Parties limit the use of Z clause.

5.3.2 Early Warnings

The early warning procedure ensures Parties undertake proactive risk management in time. Core Clause 15 of NEC4 provides the detailed early warning procedure. It is an effective approach to facilitate effective risk management. However, early warning cannot fully take over the risk management. Consequently, the terminology has changed in NEC4. For example, "risk reduction meeting" in NEC3 has been changed to "early warning meeting", and "risk register" has been changed to "early warning register."

NEC4 has added additional procedures to improve the early warnings implementation. The *Project Manager* needs to issue the first Early Warning Register within one week of the project commencement, and the first early warning meeting must be undertaken within two weeks of the project start. This assures a good starting point of the early warning process. Figure 5.4 describes the early warning process under the NEC4 contract.

Under Clause 15.1, both the *Project Manager* and the *Contractor* have the duty to notify the early warnings as soon as they are aware. Clause 15.1 provides

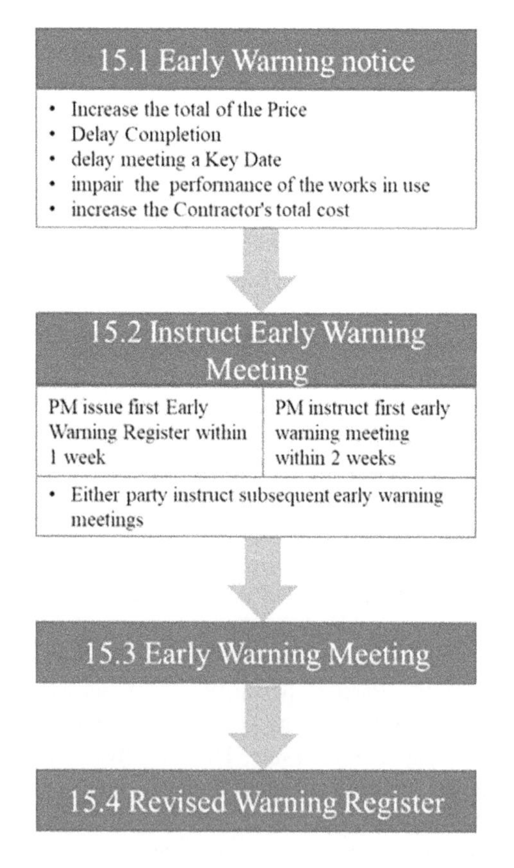

FIGURE 5.4 NEC4 early warning process.

five circumstances in which a Party needs to raise an early warning notice as follows:

1. *increase the total of the Price*
2. *delay Completion*
3. *delay a Key Date*
4. *impair the performance of the works in use*
5. *increase the Contractor's total cost*

Please note, the fifth point is not clearly listed in Clause 15.1 but is embedded in the follow-on text. Nevertheless, it is one of the most well-used grounds to notify early warning.

After the notification of the early warning, either Party can instruct the other Party and other project stakeholders to attend an early warning meeting if the other Party agrees. The mitigation options will be discussed in the early warning meeting and a decision may be made for the action to be taken. The *Project Manager* then updates the Early Warning Register following the early warning meeting and circulates to the participants.

The NEC4 contract provides potential sanctions to both the *Project Manager* and the *Contractor*, if either of them fail to perform as required by the contract. For instance, the *Client* or Others agreed to an action in the Accepted Programme, under the fourth bullet of Clause 31.2; if the *Client* or Others fail to complete the action in accordance with the Accepted Programme, the *Project Manager* should notify an early warning to the *Contractor* under Clause 15.1. If the *Project Manager* fails to notify the early warning, then the *Contractor* is entitled to a compensation event under Clause 60.1(5) as the *Client* or Others do not work within the times shown on the Accepted Programme.

On the other hand, if the *Contractor* fails to notify a compensation event for an event of which an experienced *Contractor* should have been notified as an early warning, the *Project Manager* should notify the *Contractor* that an early warning should be notified under Clause 61.5, then the compensation event will be assessed as if the early warning notice has been served under Clause 63.5. In this circumstance, the estimation of the compensation event is usually lower than the *Contractor*'s expectation. Under Clause 11.2(25) of Options C, D, E and Clause 11.2(26) of Option F, the associated cost may be evaluated as disallowed cost. This acts as a sanction for the *Contractor* not to comply with the required early warning process. Please note, the *Project Manager*'s notification under Clause 61.5 is the condition precedent to assess the compensation event as if an early waning has been notified under Clause 63.5.

Early warning notice is a successful feature of the NEC contract. As a consequence, the FIDIC 2017 edition adopts this concept and has added a new Clause 8.4 "Advanced Warning," which will be discussed further in Section 5.4.

5.3.3 Financial Benefit

Undertaking the effective risk management from the front-end of the project and continuing throughout the project life cycle would increase the success of the project which can be a win-win solution for both contract Parties. Under the Options C and D contract, the savings can be shared between the contract Parties. Under the Options A and B contract, the obvious financial benefit is in line with the *Contractor*, and can increase the profit margin. Nevertheless, the *Client* would obtain greater benefit from the early completion of the project and operation of the product.

5.4 RISK MANAGEMENT UNDER FIDIC FORMS OF CONTRACT

In the 1999 edition, risk management under FIDIC forms of contract are more focused on risk allocation between contract Parties and third-party insurance. FIDIC 2017 edition incorporates experience and lessons obtained in the 18 years of practice and introduced some of the global best practice into the contract. Advanced warning is one of the new features attributed to the enhancement of risk management and improvement of project success.

5.4.1 Risk Allocation

Similar to other standard forms of contract, FIDIC also pre-allocates some risks to the contract Parties. If the events under the Employer's Risk arise, the Contractor can then claim the Extension of Time together with additional cost and expenses under the contract. FIDIC pursues a "fair and equitable" approach to allocate risks among contract Parties. Under FIDIC 1999 edition, the defined term "Employer's Risk" and Clause 17 "Risk and Responsibility" caused some confusion to the users. Therefore, Clause 17 has revised this as "Care of the Works and Indemnities" in FIDIC 2017 edition and the defined term for "Employer's Risk" has been removed.

5.4.1.1 Employer's Risk

As discussed in Chapter 2, the Employer bears most risk under the FIDIC Red Book and bears the least risk under the FIDIC Silver Book. Table 5.4 provides the list of Employer's risk events that allows the Contractor to claim additional time and/or money under the FIDIC 2017 Red and/or Yellow Books.

These risks cover the typical Employer risk elements such as change of information provided by the Employer, including both error or scope change, unforeseen ground conditions, and *force majeure*, etc. The contract specifically sets up a process to cope with the unforeseeable physical conditions, as described in Figure 5.5. In order to obtain additional cost and time, the Contractor has to pass the unforeseeable test and prove adverse impact on the progress and cost. In addition, the Contractor needs to comply with the process under Clauses 4.12.1 to 4.12.3 as the precedent condition to variations under Clause 13.3.1 or claim under Clause 20.2. The Contractor needs to notify the Engineer of the physical conditions it encountered, give the reason for the conditions, explain why it is unforeseeable, and provide the expected impact in terms of delay of time and additional associated cost. Although there is no specific time limit for the Contractor's notification, the contract requires that the notification occurs in good time. The Engineer then needs to undertake an investigation within seven days. Similar to NEC, the Contractor is required to continue its performance during this process.

FIDIC 2017 edition also replaces Clause 19 "Force Majeure" in the 1999 edition with Clause 18 "Exceptional Events". Nevertheless, the relevant events, the procedure, and consequences are similar to the 1999 edition.

TABLE 5.4
Employer's Risk Under FIDIC 2017 Red and Yellow Contract

Clause	Description	Money	Time
1.9	Delayed Drawings or Instructions (Red) Errors in the Employer's Requirements (Yellow)	√	√
1.13	Employer's delay or failure to obtain permits, license or approval	√	√
2.1	Delayed access to the Site	√	√
4.6	An unforeseeable request to co-operate	√	
4.7	Setting out errors in items of reference	√	√
4.12	Unforeseeable Physical Conditions	√	√
4.15	Changes to access route by the Employer or a third party	√	√
4.23	Archaeological and Geological Findings	√	√
7.4	Delayed test attributable to the Employer	√	√
7.6	Remedial work attributable to the Employer	√	√
8.5(c)	Adverse climatic conditions		√
8.5(d)	Unforeseeable shortages in personnel or Goods		√
8.5(e)	Prevention by the Employer		√
8.6	Delay caused Authorities		√
8.10	Consequence of Employer's Suspension	√	√
10.2	Employer taking over part of the Works	√	√
10.3	Interference with Tests on Completion	√	√
11.7	Delay in permitting access to the Works during Defects Notification Period	√	√
11.8	Engineer instructs Contractor to search for cause of a defect which is not remedied	√	√
Y12.2	Employer's delay or interference with Tests after Completion	√	√
Y12.4	Employer delay the Contractor's access to investigate the cause of any failure to pass the Tests after Completion	√	√
13.3	Variations	√	√
13.6	Adjustments for Changes in Laws	√	√
16.1	Contractor suspends Works due to non-payment or Employer fails to comply with the final & binding decision by Engineer or DAAB.	√	√
16.2.2	Delay of notice period to terminate the contract due to Employer's default	√	√
16.3	Additional works instructed by Engineer to protect life, property or the safety of the Works following termination	√	√
17.2	Loss or damage to the Works caused by Employer's risk event	√	√
18.4	Consequences of an Exceptional Event	√	√

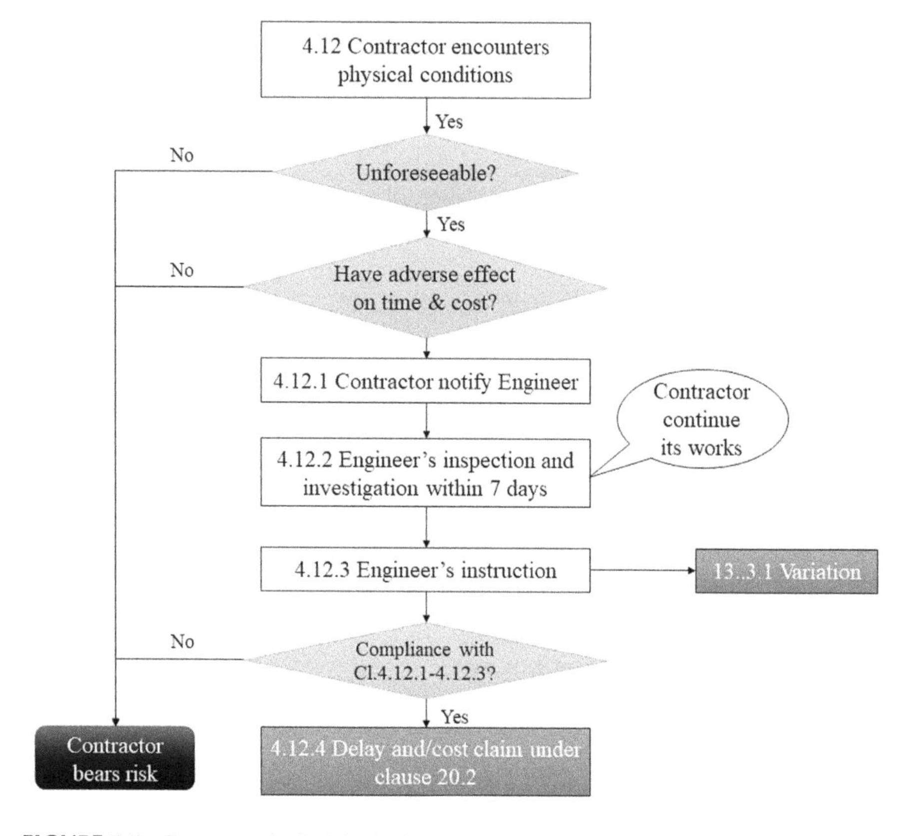

FIGURE 5.5 Process to deal with physical conditions under FIDIC 2017.

FIDIC 2017 edition provides some new Employer risks, which give the Contractor claims to additional time and money, including the following:

- Clause 1.13 Employer's delay or failure to obtain permits
- Clause 4.15 Changes to access route
- Clause 7.6 Remedial work attributable to the Employer
- Clause 16.2.2 Notice period prior to termination for Employer default

Under the Silver Book (FIDIC, 2017), the Employer does not bear risk of use or occupation by any part of the Permanent Works and unforeseeable operation of forces as in the Red Book and the Yellow Book. In addition, because the Silver Book is a Turnkey contract, the Contractor is liable for the design even if the design is undertaken by the Employer's personnel.

5.4.1.2 Contractor's Risk

In large construction projects, the typical Contractor risks include:

- Time: delay
- Cost: overrun

- Quality: defects associated with both patent defects and latent defects
- Loss and damages during the period of implementation
- Change of exchange rate or inflation rate
- Sub-contractor
- Obtaining sufficient material and plant
- Construction methodology

Under the FIDIC Yellow Book and Silver Book, the Contractor also bears design risks associated with design change, design mistake, and design liability.

In addition, the Contractor bears risk of their poor management, e.g., failure to procure sufficient material required on time, applying inappropriate construction methodology, and failure to comply with the contract's procedure. For example, the Contractor loses its entitlement to claim for the Employer's risk event as a consequence of time bar due to the lack of competent knowledge or skill of their managers.

In addition, the Parties may also transfer some risks to the third party via insurances. For example, the consultant's professional indemnity insurance transfer risks associated with its design liability to the insurance company.

5.4.1.3 Exceptional Events

Clause 18 "Exceptional Events" relates to the "Force Majeure" in FIDIC 1999 edition, it is also relevant for the event under Clause 60.1.(19) of the NEC contract. Clause 18.1 provides the definition of "Expressional Event" as the event:

(i) *is beyond a Party's control;*
(ii) *the Party could not reasonably have provided against before entering into the Contract;*
(iii) *having arisen, such Party could not reasonably have avoided or overcome; and*
(iv) *is not substantially attributable to the other Party.*

It then provides the example of events that can be categorized as an "Expressional Event" as follows:

(a) *war, hostilities (whether war be declared or not), invasion, act of foreign enemies;*
(b) *rebellion, terrorism, revolution, insurrection, military or usurped power, or civil war;*
(c) *riot, commotion or disorder by persons other than the Contractor's Personnel and other employees of the Contractor and Subcontractors; (d) strike or lockout not solely involving the Contractor's Personnel and other employees of the Contractor and Subcontractors;*
(e) *encountering munitions of war, explosive materials, ionising radiation or contamination by radio-activity, except as may be attributable to the Contractor's use of such munitions, explosives, radiation or radio-activity; or*
(f) *natural catastrophes such as earthquake, tsunami, volcanic activity, hurricane or typhoon.*

In order to obtain additional time and cost for the impact of the "Expressional Event," the Contractor needs to follow the specific process as shown in Figure 5.6. The affected Party needs to serve the first notice to the other Party of the prevention

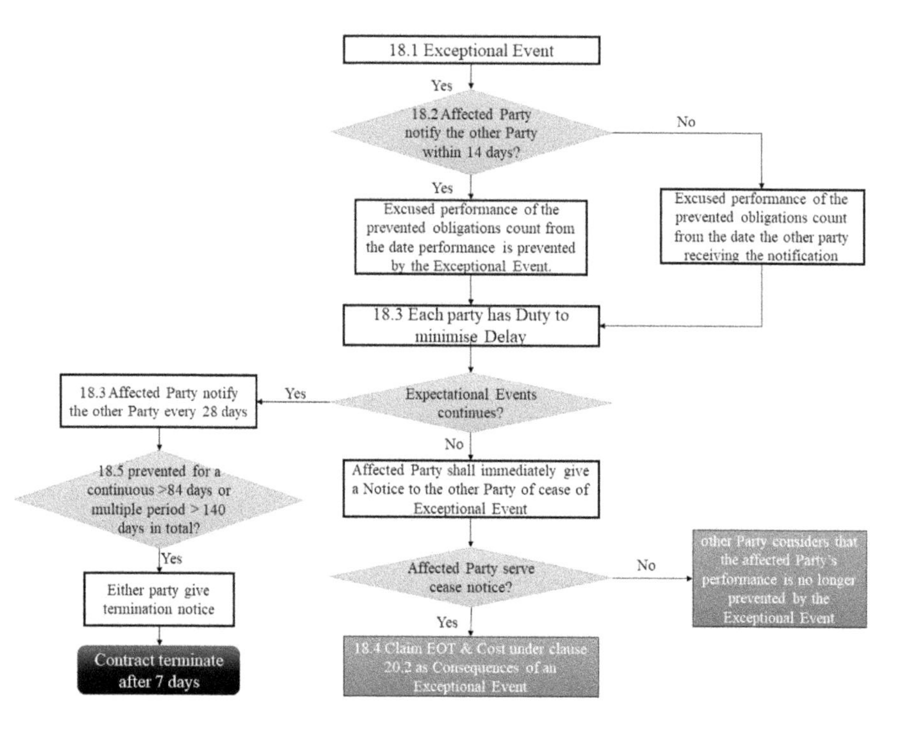

FIGURE 5.6 "Expressional Event" process under FIDIC 2017 contract.

of the "Expressional Event" within 14 days, then the performance of prevention will be counted from the date that the "Expressional Event" occurs. Otherwise the prevention will be counted from the date of the other Party receiving the first notification. This minor sanction intends to incentivize the Parties to notify the other Party of the "Expressional Event" in time. Both Parties then have the duty to minimize the delay. The affected Party must notify the other Party immediately when the "Expressional Event" has stopped. Failure to serve the cease notification will result in the affected Party losing the compensation for the prevention, as the other Party will consider that the performance was not prevented by the "Expressional Event." This sanction provides another incentive for the Parties to follow the process set out in the contract. If the "Expressional Event" continues, the affected Party needs to notify the other Party every 28 days, until the "Expressional Event" stops. However, if the Expressional Event continues for 84 days or the total of the multiple prevented period is more than 140 days, then either Party can serve a termination notice to the other Party. The contract is then terminated seven days after the termination notice.

5.4.2 Advanced Warning

FIDIC 2017 edition provides a new provision for "Advanced Warning" under Clause 8.4. It requires the Contractor or the Employer to serve the advanced warning to the other Party and the Engineer for the events in the following four circumstances:

(a) *adversely affect the work of the Contractor's Personnel;*
(b) *adversely affect the performance of the Works when completed;*
(c) *increase the Contract Price; and/or*
(d) *delay the execution of the Works or a Section (if any).*

It further provides that "*The Engineer may request the Contractor to submit a proposal under Sub-Clause 13.3.2 [Variation by Request for Proposal] to avoid or minimise the effects of such event(s) or circumstance(s).*"

Unlike the systematic early warning process in NEC, Sub-Clause 8.4 in the FIDC 2017 forms of contract only extends to potential "Variation by Request for Proposal" under Sub-Clause 13.3.2. There is no time limit for such advanced warning or sanction for the Party that does not serve an advanced warning. Although it is good to introduce this global best practice approach into the FIDIC contract, it may not achieve the same success as NEC under the current set-up.

5.5 TIME RISK MANAGEMENT

5.5.1 Schedule Risk

In Chapter 3, the important of time management was discussed as it can have a major impact on the delay damages. Therefore, understanding and managing schedule risk is important to large construction projects. Due to the complexity of the programme for large construction projects, the schedule risk management is different from the normal methods undertaken in general project management. Together with the general development of information technology, several pieces of software have been developed in particular to undertake comprehensive quantitative schedule risk analysis, which is often known as QSRA. PertMaster was the first well-known software that undertook QSRA; it is integrated with Primavera Project Planning (P3). Oracle then acquired both P3 and PertMaster and further developed Primavera Enterprise Project Portfolio Management (P6) and Primavera Risk Analysis (PRA) respectively. With the development of its own cloud and data lake platform, Oracle then integrated the key project management packages, including schedule, risk, lean, document control, portfolio planning into a cloud-based "PRIME." Furthermore, Deltek developed Acumen Risk which provides most of the functions that PRA does, and have added additional features, e.g., a more effective Tornado chart. In recent years, Safran Risk also gave more attention to QSRA with a competitive price and the special features to view the impact of any combination of risk throughout the project quickly.

5.5.2 QSRA (Quantitative Schedule Risk Analysis)

Because QSRA relies on the relationship and duration of the existing programme, the accuracy of the relationship between the activities and estimated durations can have a substantial impact on the output of the analysis. Therefore, it is important to carry out schedule quality analysis before undertaking QSRA.

All QSRA software provide the function of schedule quality check and provide a report of warnings or errors found, including the open-ended activities, activity with constraints, broken logic, activities with negative lags, start to finish relationships, lags between tasks with different calendars, etc. The planner needs to fix all the open-ended and broken logic errors and as much as possible other warnings before undertaking QSRA. It is also recommended to remove all constraints before carrying out QSRA.

5.5.2.1 Duration Uncertainty

There are two components that contribute to the quantitative schedule risk analysis: the duration uncertainty and the impact of risks listed in the risk register. The duration uncertainty concerns the accuracy of the duration estimated during the initial project planning, for instance the Time Risk Allowance (TRA) under the NEC contract can be considered as part of the duration uncertainty. The duration uncertainty is usually defined with a three-point estimate, which defines the minimum duration, most likely duration, and maximum duration for each activity. The risk analysis software will allocate a pre-defined distribution for each activity defined with the three-point duration values. The commonly used statistic distribution is triangular distribution.

With large construction projects, it would be a time-consuming job to define the duration uncertainty for thousands of activities. PRA allows the user to define duration uncertainty quickly with the percentage range from the most likely value together with the distribution profile. The user can select the activities in the programme, use a pre-defined filter to select groups of tasks, or simply apply the general principle to all activities in the programme. For the selected group of tasks, the user can define the minimum duration, most likely duration, and maximum duration in percentage of the existing remaining duration in the programme, for example, 80% for minimum duration, 100% for most likely duration, and 115% for the maximum duration. The user can also select the different distribution for the different activities, including Triangle, BetaPert, Enhanced, Uniform, Normal, modified BetaPert, Cumulative, general, Trigen, Lognormal, and Discrete distributions. In PRA, the user can choose from four types of distributions, including uniform distribution, triangle distribution, BetaPert distribution, and Enhanced distribution. The most well-used distributions are triangle and BetaPert distributions.

5.5.2.1.1 Uniform Distribution

If a certain activity has quite a fixed duration, and needs hardly any changes, then the uniform distribution can be used. Under uniform distribution, the minimum duration, the most likely duration, and the maximum duration are all the same.

5.5.2.1.2 Triangle Distribution

Triangle distribution is the often-used choice in practice. Equation 5.2 describes the calculation of triangular distribution.

$$\text{Activity Duration}_{\text{Triangular}} = \frac{D_p + D_m + D_o}{3}$$

Equation 5.2 Three-Point Duration Estimate with Triangular Distribution

As shown in Figure 5.7, the distribution value is in a linear relationship between the most likely value to the minimum value, and between the most likely value to the maximum value.

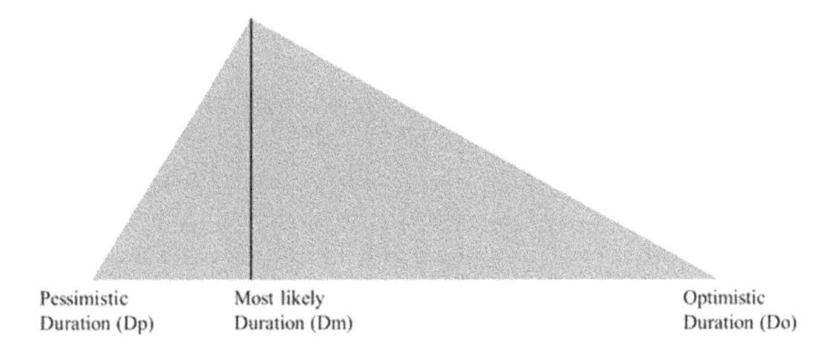

FIGURE 5.7 Three-point duration estimate.

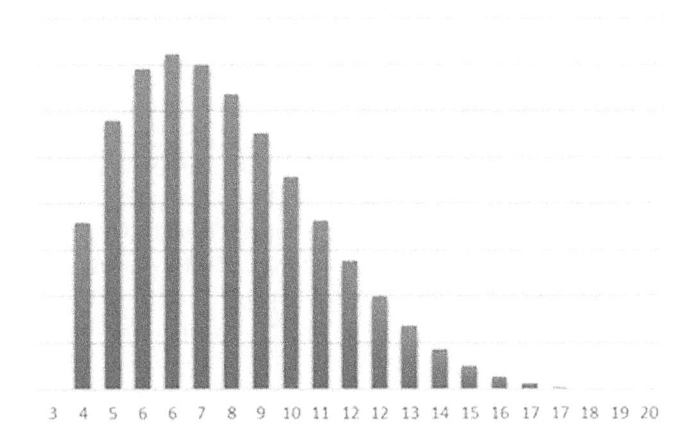

FIGURE 5.8 BetaPert distribution for activity duration.

5.5.2.1.3 BetaPert Distribution

BetaPert distribution is another well-used distribution when undertaking QSRA. Like Triangle distribution, BetaPert distribution also measures three points: the minimum duration, the most likely duration, and the maximum duration, but it smooths the curve and demonstrates more confidence with the most likely duration, as shown in Figure 5.8.

Equation 5.3 describes the calculation of the BetaPert distribution.

$$\text{Activity Duration}_{\text{Beta}} = \frac{D_p + 4D_m + D_o}{6}$$

Equation 5.3 Three-Point Duration Estimate with BetaPert Distribution

5.5.2.1.4 Enhanced Distribution

Enhanced distribution creates a four-point distribution from optimistic, most likely, and pessimistic. Then the extended minimum and extended maximum are calculated as follows:

$$\text{Extended Duration}_{\text{min}} = D_p - 0.6 \times (D_o - D_p)$$

Equation 5.4 Enhanced Distribution Extended Minimum Duration

$$\text{Extended Duration}_{max} = D_o + 2 \times (D_o - D_p)$$

Equation 5.5 Enhanced Distribution Extended Maximum Duration

The extended minimum and maximum duration are used to simulate the possibility of the actual duration exceeding the defined minimum and maximum duration. It is used when considering that the actual duration may not approach the defined extreme duration for each end. Figure 5.9 explains the enhanced distribution for an activity with 12 days duration and minimum 75%, most likely 100%, and maximum 125% definition.

For large construction projects, often the programme contains over 10,000 activities and it would be very time consuming to define the duration uncertainty for each individual activity when undertaking schedule risk analysis. It would be handy to set up a duration uncertainty template for the whole project by grouping together activities with similar features. This template can be defined by disciplines, phase, work packages, or types of works which share similar patterns of duration uncertainty. For example, to allocate excavation in one code and allow 95% as the minimum duration, 100% as the most likely duration, and 150% as the maximum duration; and allocate brick works for another code, with 90% as the minimum duration and 110% as the maximum duration.

Meanwhile, the project control manager or planning manager should set up a user defined field in the Primavera P6 and request each planner to allocate a duration uncertainty code for each activity. Table 5.5 demonstrates an example of using the duration uncertainty template for Primavera Risk Analysis. The activities will be coded in several groups for each discipline. For instance, Civils works will be grouped into four categories and the activities with similar duration features will be allocated a code between C1 to C4. Similarly, the code allocation will be applied to mechanical, electrical, fit out, testing, and commissioning activities. After importing the P6 file into the PRA, and simply applying the duration uncertainty template, the duration uncertainty definition for each activity can be completed for the whole

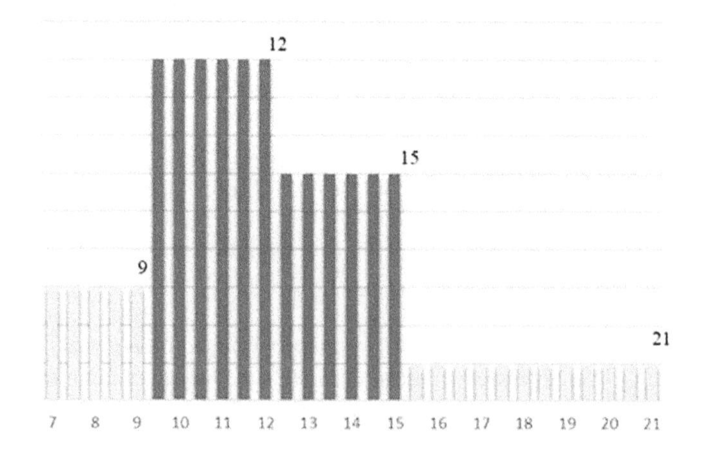

FIGURE 5.9 Enhanced distribution for activity duration.

TABLE 5.5
Duration Uncertainty Template for Primavera Risk Analysis

DU Code	Distribution	Min.	Likely	Max.
C1	Triangle	70%	100%	110%
C2	Triangle	75%	100%	125%
C3	BetaPert	95%	100%	105%
C4	Enhanced	80%	100%	130%
E1	Triangle	85%	100%	120%
E2	Triangle	75%	100%	125%
F1	Enhanced	90%	100%	115%
F2	Triangle	80%	100%	120%
M1	Enhanced	75%	100%	130%
M2	Triangle	85%	100%	110%
T1	Triangle	80%	100%	125%

project simply by applying the pre-defined template. The risk manager can then conduct a review and make necessary adjustments to the duration uncertainty values.

For projects under the NEC contract, Clause 32.1 requires the programme to show the Time Risk Allowance (TRA). The TRA is usually shown in a user defined column in the P6 programme. In this circumstance, it would be handy to use the exiting TRA to define the duration uncertainty with the enhanced or BetaPert distribution and then adjust the lower and upper end percentage.

5.5.2.2 Impact of Risks

The second stage of QSRA is risk mapping. Primavera Risk Analysis (PRA) provides an embedded risk register function. The risks listed in the risk register can then be allocated to the activities that may have an impact. Meanwhile, PRA can create a new model which inserts the impacted risks as separate activities integrated into the existing plan, as shown in Figure 5.10.

PRA can also generate the plan which is integrated with risk mapping for pre-mitigation scenarios and post-mitigation scenarios respectively. After integrating the risks into the programme, the Monte Carlo simulation is then undertaken with the risk model assessing both the duration impact for each activity to the schedule networks and the impact of the risks to the relevant activities as well as the whole project. The output of QSRA includes:

- the Tornado chart to indicate the sensitivity of the top risks as well as top activities, which have major impact to the whole project completion date
- the confidence level of achieving each activity, key milestone, and the project completion date
- exporting the critical path of the simulation results, which take into account the duration uncertainty and impact of key risks

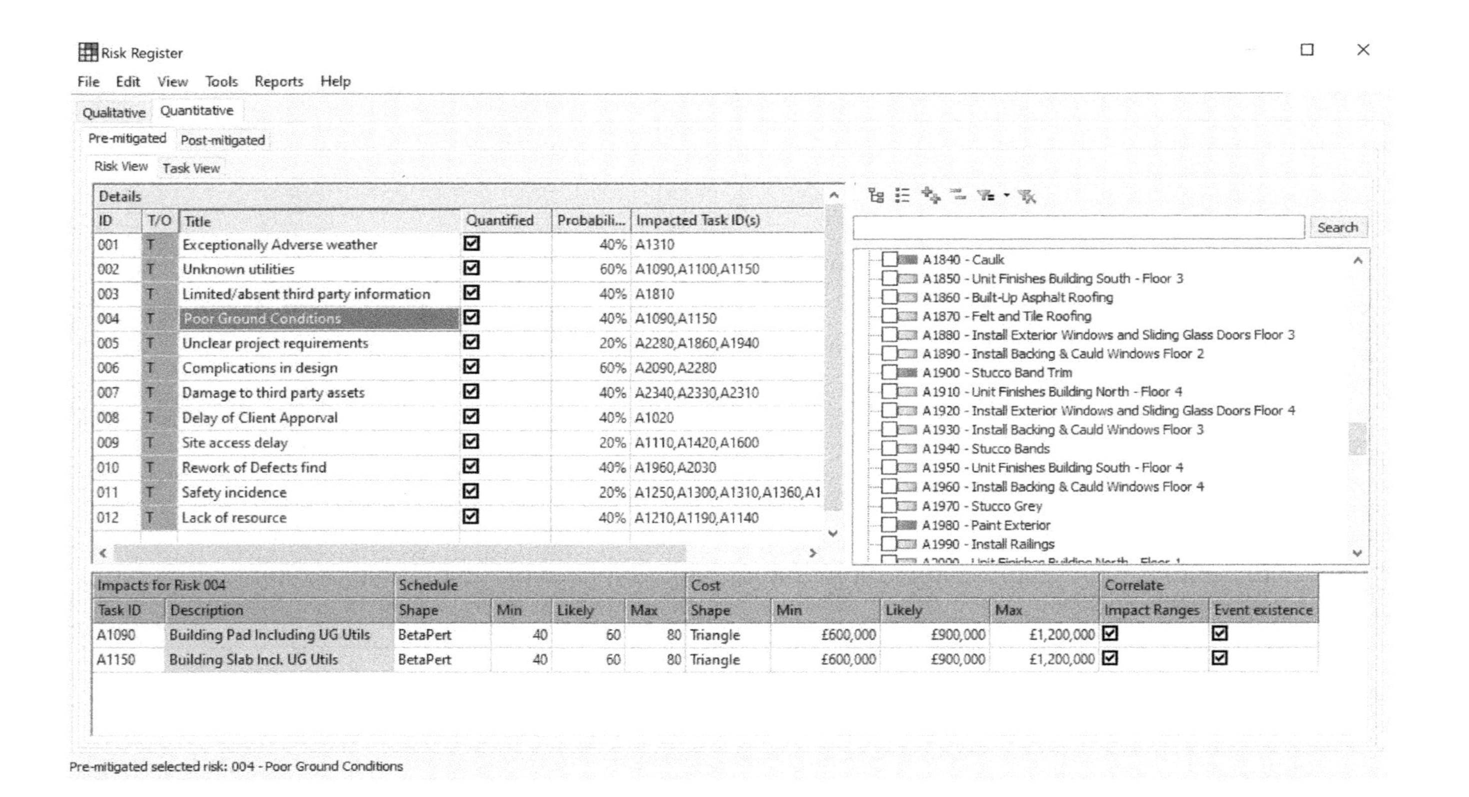

FIGURE 5.10 Risk mapping to programme activities in Primavera Risk Analysis. (Copyright Oracle and its affiliates. Used with permission.)

5.5.2.2.1 Confidence Value for Completion

In practice, the P80 dates are generally used to determine the realistic completion date. Figure 5.11 demonstrates an output of the QSRA result. It illustrates that, under the existing scenario, there is only a 19% chance of completing the project by the contract completion date, which is 5th February 2021. There is a 50% chance that the project can be completed by 2nd March 2021, and an 80% change that the project can be completed by 25th March 2021.

Based on the P80 date, the project manager would expect about seven weeks' delay under the existing circumstances. The project management team should consider the key risks driving the programme as well as the key activities that have been impacted to mitigate the potential delay of the project.

5.5.2.2.2 Tornado Chart

With Tornado charts, the key risks that have a major impact on the project can be identified. For example, Figure 5.12 provides the top activities that are likely to impact the project completion date.

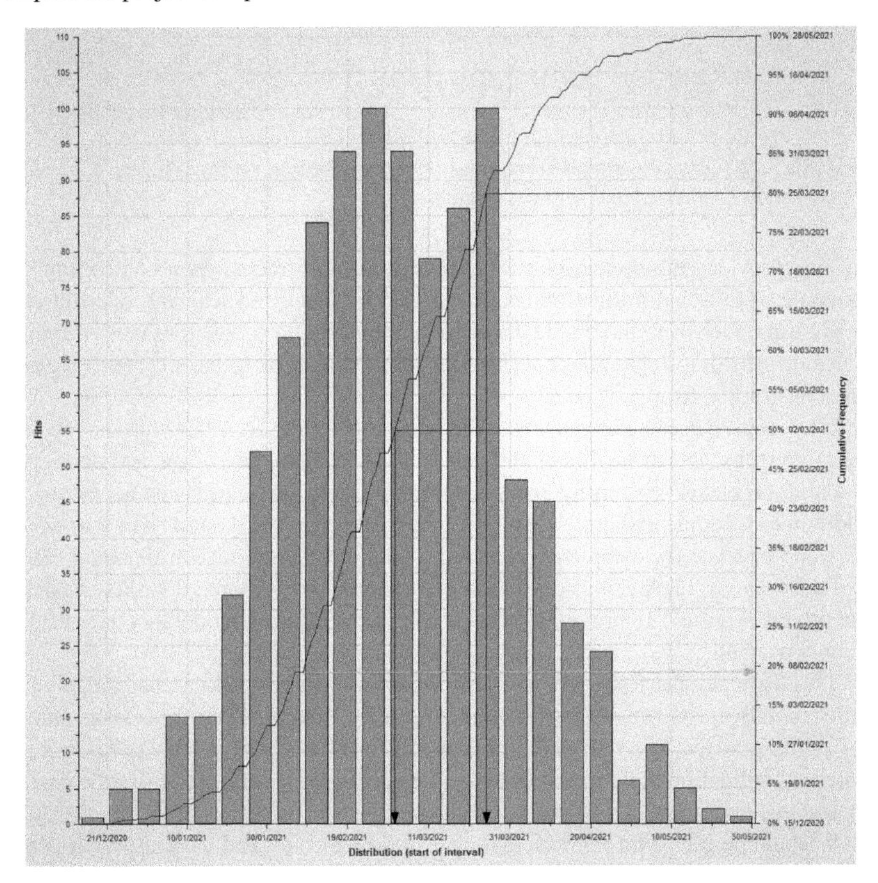

FIGURE 5.11 Schedule risk analysis output in Primavera Risk Analysis. (Copyright Oracle and its affiliates. Used with permission.)

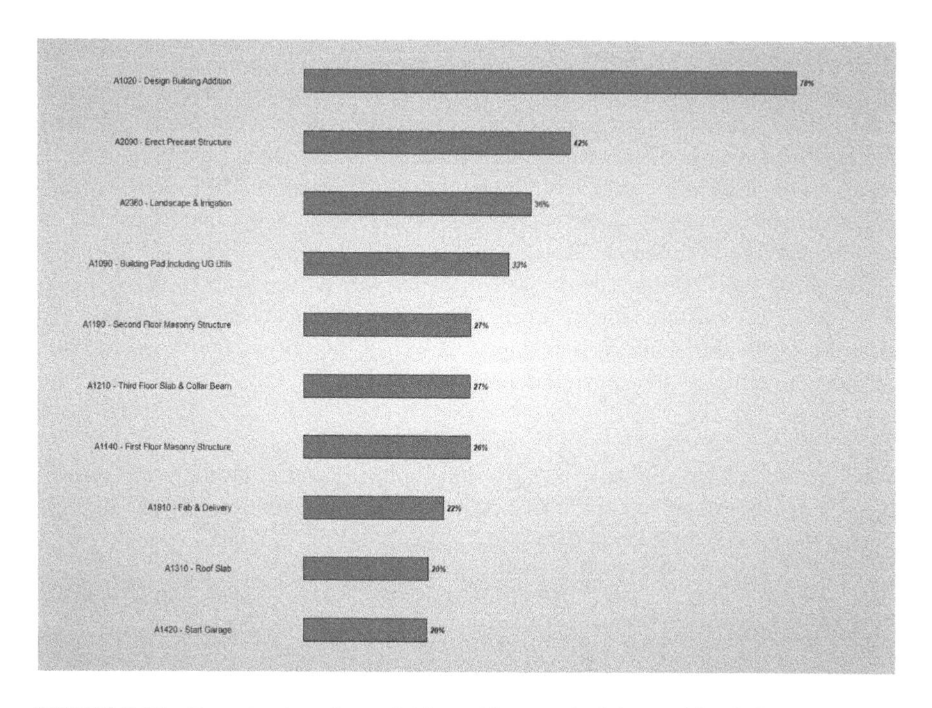

FIGURE 5.12 Tornado chart for activities with top schedule sensitive index. (Copyright Oracle and its affiliates. Used with permission.)

This provides the project control manager areas to focus when controlling the schedule. Additional mitigation actions should be discussed with the relevant parties in time, such as reducing the duration by increasing resource for tasks with high schedule sensitivity, changing construction method or work sequence, resolving certain constraints, etc.

Likewise, PRA can also provide a Tornado chart for the activities with the top cost sensitive index, which facilitates the cost control team to control the activities with potential problems proactively and raise additional change requests as necessary. In addition to schedule and cost sensitivity, a Tornado chart for the top risks that would have an impact on the completion date and total project cost is also available in PRA.

Furthermore, PRA can also provide a Tornado chart for top risks likely to delay the project. Figure 5.13 provides the top risks in the existing model that may lead to project overrun.

This Tornado chart provides the risk manager with key risks to be mitigated in order to deliver the project in time and within the budget. Then these risks can be set as top priority. The risk manager can call for all relevant parties to discuss the potential mitigation options and delegate the agreed mitigation actions to the parties most competent to resolve the relevant risks.

It would be better for the risk manager to take a combined view of the Tornado chart for top activities as well as top risks that would cause the cost overrun and/or project delay. The combined analysis of different type of Tornado charts can provide the focus area for both risk and activities. Consequently, the overall impact to the

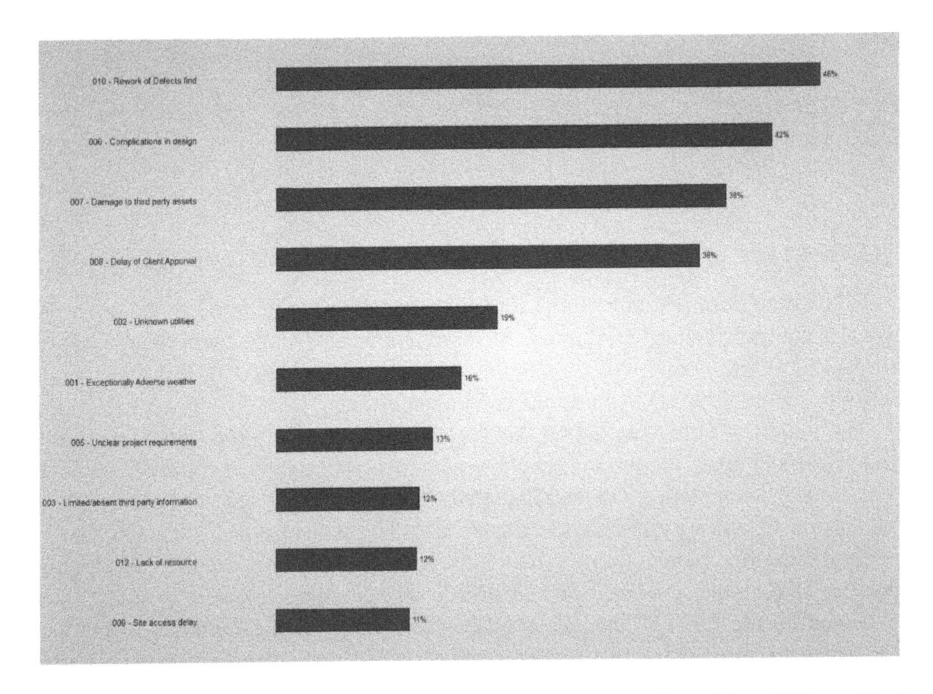

FIGURE 5.13 Tornado chart for top schedule. (Copyright Oracle and its affiliates. Used with permission.)

project delivery can be better understood, which enables the project management team to develop more effective mitigation actions to resolve the problem.

5.5.2.3 Programme Confidence

The unrealistic baseline programme is often a key reason for a project's failure to complete on time. Clause 31.3 of the NEC contract provides that the *Project Manager* cannot accept the *Contractor*'s programme, if the programme is unrealistic.

When assessing the *Contractor*'s programme, the *Project Manager* can undertake a preliminary schedule risk analysis to understand the confidence level of the *Contractor*'s programme to achieve the Completion Date. On the other hand, the *Contractor* can also use QSRA to analyze the programme before submitting it to the *Project Manager*. An Accepted Programme is important to both Parties under the NEC contract, as it resets the baseline programme of the project. A common issue with large-scale construction projects under the NEC3 contract is the lack of Accepted Programme in time. It happens quite often that for over six months or even a year, there is no Accepted Programme from the *Project Manager*. Although NEC4 introduces the deemed acceptance to incentivize both Parties to performance proactively, it is still an overall risk to the project that an Accepted Programme is not in fact realistic.

QSRA can provide clear evidence of confidence to both the *Contractor* and the *Project Manager* of how realistic the programmes. It not only fundamentally resolves issues associated with the programme acceptance, but also improves the programme delivery.

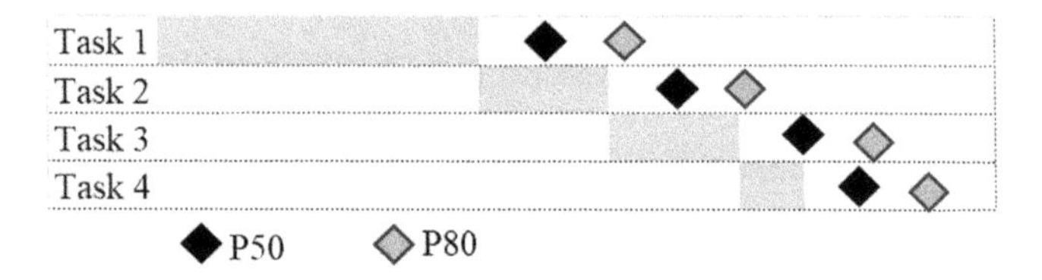

FIGURE 5.14 Programme activity with completion date at P50 and P80 confidence level.

When both duration uncertainty and risk allocation are inputted to the quantitative risk analysis model, the Monte Carlo simulation can then be undertaken. Usually at least 1,000 iterations will be simulated. For programmes with less than 3,000 activities, more than 3,000 iterations can be undertaken.

The output of QSRA including the Tornado chart for both top critical activities and top risks impact the project schedule and cost respectively, the confidence level to achieve the completion date and P50/P80 completion date for the overall project as well as for each activity as shown in Figure 5.14. The criticality index of each activity under the existing risk model, and the risk scatters that demonstrate the P50 and P80 position of both the time and cost risk profile are also impacted.

Based on the P50/P80 date for the key date and/or completion date, the Project Manager can then make the decision on whether the Contractor's programme is realistic.

5.5.3 Critical Path and What-if Scenarios

After the Monte Carlo simulation with the defined duration uncertainty and risk profile, the criticality of each activity can be reported by the PRA. The user can export the critical path report and review the critical path and sub-critical path in the programme.

For risks with a potential critical impact on the project, different what-if scenarios can be modeled with different risk mitigation actions, or construction sequences. Those scenarios can then be compared with the impact on the key milestones and the key dates as well as the total cost to undertake the risk mitigation. For example, Figure 5.15 demonstrates the distribution analysis for a programme under different scenarios.

The analysis results can then be used for a further decision tree analysis and to facilitate the project manager or senior management to select the most appropriate mitigation actions in order to provide the best result for the project.

The risk manager can also use the critical path analysis information to review and adjust the programme and provide a more realistic programme. If the completion date is likely to be delayed significantly, then the senior management needs to make an appropriate decision early and notify the key stakeholders if necessary.

5.6 COST RISK MANAGEMENT

Cost risk management is equally as important as schedule risk management. Apart from undertaking it when integrated with schedule analysis within PRA, if the schedule is resource and cost loaded. In the projects does not requires quantitative schedule risk analysis, but only requires quantitative cost risk analysis, the

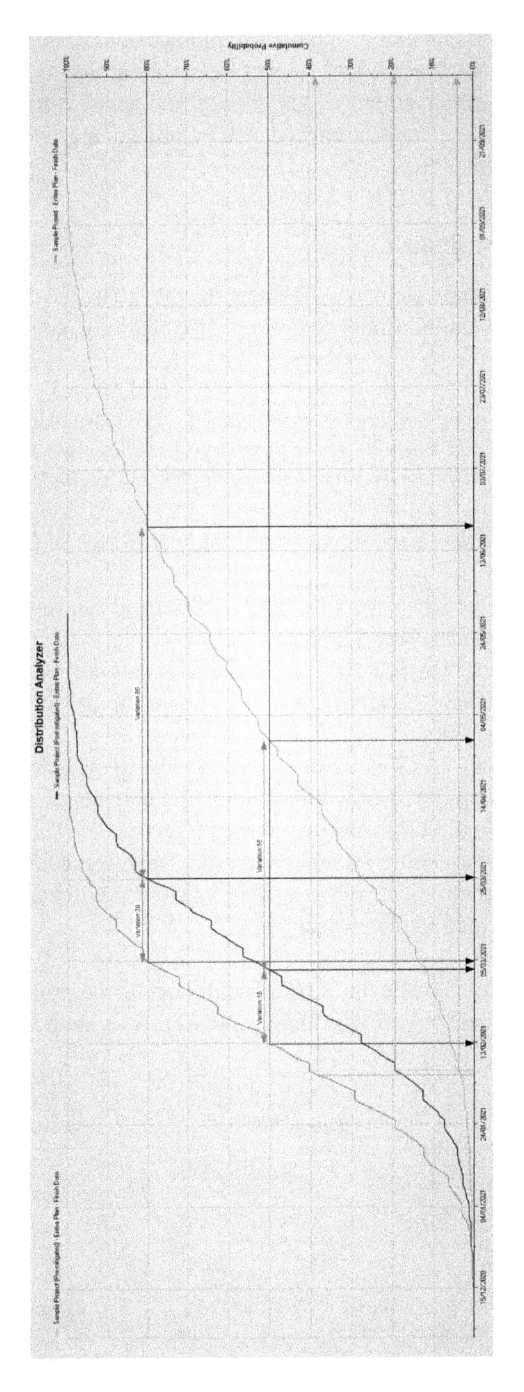

FIGURE 5.15 Distribution analysis in Oracle Primavera Risk Analysis. (Copyright Oracle and its affiliates. Used with permission.)

analysis can also be undertaken based on the risk register with the comprehensive spreadsheet model or other risk analysis software, e.g., Palisade's @RISK. Similar to time risk management, cost risk management also needs to manage the cost uncertainty, the thread, and opportunities identified in the risk register. In addition, cost risk management needs to take into account the cost associated to the schedule variance from the earned value management, which will be discussed further in Chapter 8.

5.6.1 Types of Cost Reserve

When discussing risk management, the most-quoted phase is the statement that Donald Rumsfeld, the United States Secretary of Defense, gave in a news briefing in 2002:

> *there are known knowns; there are things we know we know. We also know there are known unknowns; that is to say we know there are some things we do not know. But there are also unknown unknowns—the ones we don't know we don't know.*

PMI (2019) classified risks in accordance with four categories as follows:

- Known-Known, which is the facts and requirements set up in the project scope and do not constitute as risk.
- Known-Unknown, which is the classic risk. The project team have knowledge and experience to identify it and estimate the potential impact and occurrence probability.
- Unknown-Known, which is the hidden fact. The relevant knowledge may be available among the project stakeholders but may not be aware or available to the people directly working on the project.
- Unknown-Unknown, which is emergent risk. The relevant knowledge does not exist within the project. When such risk arises, it will certainly become a new lesson learned to the project team.

The total project budget is made up of three components: the project base cost estimate, the contingency reserve, and the management reserve, as shown in Figure 5.16.

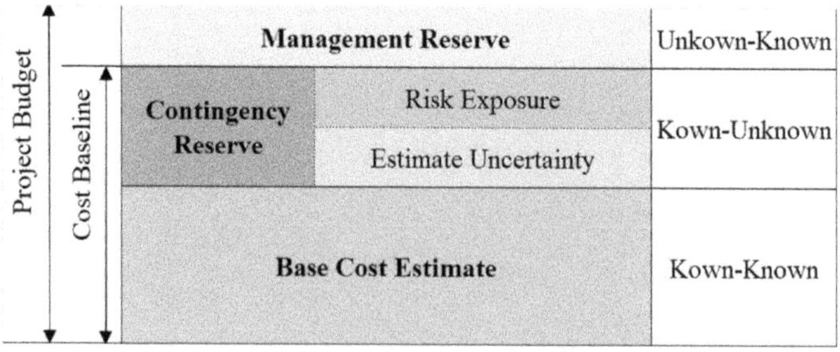

FIGURE 5.16 Types of contingency constituting project budget.

Therefore, the formula is used to calculate the project budget as show in Equation 5.6.

$$\text{Project Budget} = \text{Cost Estimate} + \text{Contingency Reserve} + \text{Management Reserve}$$

Equation 5.6 Project Budget

5.6.1.1 Contingency Reserve

The base cost estimation is built up with the sum of each project cost components. Due to the bias and uncertainty associated with the cost estimation, the project management team will usually include a sum of contingency on top of the base cost estimation to establish the cost baseline, which will then be used for the earned value management during the project progress. The contingency research can then be classified as estimate uncertainty and risk exposure. Estimate uncertainty depends on the method used to undertake the cost estimation. Generally, the bottom-up estimation method results in higher value than the top-down estimation method. During the competitive tendering stage, in order to win the bid, the senior manager usually estimate a cost range, which is likely to win the project and then distribute the budget to the work packages or disciplines. However, the senior manager may underestimate some specific components that may differ from the general practice, hence will cost more. Therefore, the estimate uncertainty will be higher in the top-down estimation.

During cost estimation and project planning, assumptions are made for the scenarios in consideration. Meanwhile the risks associated with the project are also captured. Based on the experience of the project team and the complexity of the project, the preliminary risk exposure will be estimated and the contingency to cover the existing risk exposures will be included in the contingency reserve. The contingency reserve can often be calculated with the Expected Monetary Value (EMV), as described in Equation 5.1 Expected Monetary Value, and Decision Tree analysis.

5.6.1.2 Management Reserve

Management reserve is the reserve deal with the Unknown-Known hidden fact that may arise outside the existing scope and risk exposures. Management reserve is used for the purpose of control and management of the overall project instead of completing any specific tasks required within the project. Ideally the management reserve would not be triggered at project completion. However, in adverse circumstances the management reserve will be used as part of the project funding, before deducting the project profit and realizing actual project deficit.

5.6.2 QCRA (Quantitative Cost Risk Analysis)

Due to the complexity and scale of large construction projects, it would be beneficial to use advanced computer software to assist the quantitative cost risk analysis. There are many ways that the quantitative risk analysis can support effective cost risk management.

As discussed in the previous section, QCRA can be used for determining the contingency reserve at the front-end of the project.

Similar to QSRA, QCRA also considers the cost uncertainty, likelihood, and severity captured in the risk register. In the circumstances that only quantitative cost risk analysis is required, Palisade's @RISK is usually used to undertaken QCRA based on the risk register. Like duration uncertainty in QSRA, for each risk item in the risk register, the cost uncertainty is defined with minimum cost, most likely cost, and maximum cost. The cost distribution profile is then defined based on the characteristics of the individual risk item and range of cost uncertainty. In addition, the occurrence probability can be calculated based on the likelihood and the severity defined with the typical risk matrix as shown in in Figure 5.1. The distribution profile for both likelihood and severity can also be defined based on the features of individual risk item. Following the definition of the distribution profiles, the Monte Carlo simulation can be undertaken.

Figure 5.17 explains the components of the quantitative cost risk analysis. Please note the cost uncertainty here is different from the estimate uncertainty that contributes to the contingency reserve discussed in the previous section. The cost uncertainty here is more related to the elements that are not sufficient to escalate as risks to map to the programme activities in the quantitative cost risk analysis.

The total of the cost uncertainty, thread, opportunity, and schedule risk will then conclude the total cost risk. Following the Monte Carlo simulation, the deterministic mean value, confidence P values (e.g., P50, P80, or P90) can be obtained from the quantitative cost risk analysis result. The total of these four components then provide the overall picture of the cost risk exposures to the project manager, client, and the project board and support any further decisions. The outcome of the quantitative cost risk profile provides the information to facilitate decision making of the contingency reserve.

For large construction projects, it would be better to use quantitative risk analysis software, e.g., Palisade's @RISK or Primavera Risk Analysis to decide the contingency risk level.

First, the base cost estimate has been evaluated with the estimate uncertainty. Second, apply the existing known risk components to the project and the confidence level of the project outturn cost will be generated as a result of the quantitative cost risk analysis; it is common to use the P80 or P90 value in conjunction with the

	Mean Deterministic	**P50**	**P80**
Cost Uncertainty			
Thread			
Opportunity			
Schedule Risk			
Total Cost Risk			

FIGURE 5.17 Quantitative cost risk analysis component.

estimate uncertainty to determine the contingency reserve of the project. If the P80 project cost confidence level is in line with the range of base cost plus the contingency, then the chance of the project being completed within the budgeted cost is high. If the P80 ends within the range of management reserve, then the project is running with the risk of cost overrun. If the P80 confidence level is beyond the management reserve, then the organization undertaking the project is running with risks to the business.

This approach will provide the project leader with a more accurate contingency reserve and specific confidence to manage the project in the future.

Similar to QSRA, the output of QCRA also provide the Tornado chart for top risks that has significant impact to the whole project cost. The project manager or risk manager can then priories the effort to mitigate these top risk items in order to effectively mitigate the cost risk of the overall project.

5.6.3 QSCRA (Quantitative Schedule and Cost Risk Analysis)

In practice, it is more efficient to integrate both quantitative schedule risk analysis and quantitative cost risk analysis together. When undertaken QSRA, if the programme is cost and resource loaded, the QCRA can also be undertaken simultaneously. As discussed in section 5.6.2, when undertaken QCRA, due to the delay of completion triggering the delay damages, the cost impact associated with the schedule risk should also be considered in the overall quantitative cost risk analysis. As a result of the integrated quantitative schedule and cost risk analysis (QSCRA), the results can be reviewed jointly to evaluate the overall impact to the project. For example, Figure 5.18 shows the scatter chart provided by Primavera Risk Analysis after the joint QSCRA is complete.

The trend of the schedule and cost risk profile can also illustrate whether the existing risk profile has a high impact on schedule or cost. If the slope is higher than the cashflow profile of the overall project, then it illustrates that the cost risk is higher than the schedule risk, and vice versa.

5.6.4 Risk Drawdown

As discussed in Section 5.6.1, the contingency reserve will be retained separately for the project. For large construction projects, it is important to establish a procedure to draw down the reserved contingencies at the front-end of the project.

This contingency will be managed by the project manager, commercial manager, or the project control manager. It is important to understand what constitutes the initial contingency reserve, how much relates to the initial estimate uncertainty, and how much relates to the initial risk exposure. In order to control the risk drawdown appropriately against the initial contingency set-up, the risk status for the initial identified risks needs to be updated. A dashboard report can be set up to report the risk drawdown status for each period. Figure 5.19 provides an example of the project risk drawn down dashboard report.

The budgeted contingency can be drawn down in accordance with the cost overrun in the period and the estimate uncertainty. The risk profile depends on when the

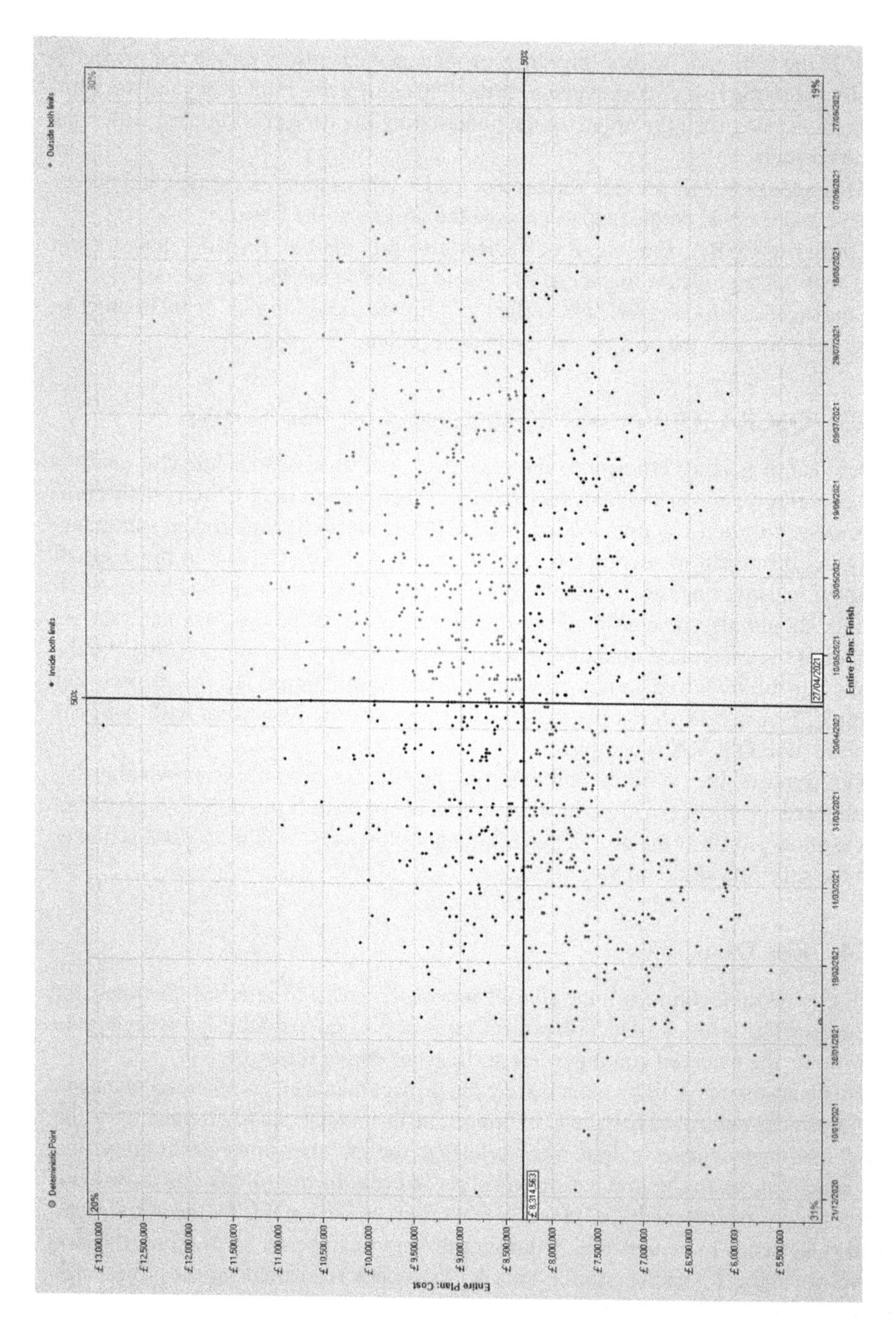

FIGURE 5.18 Scatter chart for schedule and cost risk profile. (Copyright Oracle and its affiliates. Used with permission.)

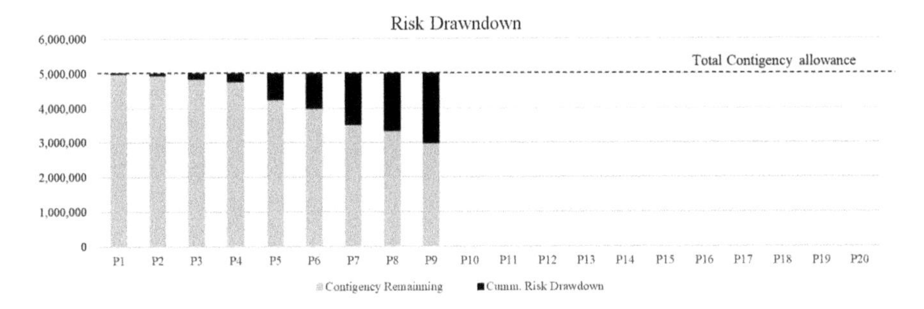

FIGURE 5.19 Risk drawn down dash board report.

risk will come into effect, the duration impacted, and the impacted value. Similar to the earned value management, the status for the components of the contingency reserve will then be updated in the same interval of the project payment. If the actual cost is within the estimation range, then the estimate uncertainty can be saved for later use. Likewise, if the initial risk has been mitigated with less cost allowed in the contingency reserve, then the savings can be carried over for the rest of the project. However, if the actual cost is higher than the contingency reserve, then the risk allowance will turn to negative, which will warn the project manager to take appropriate mitigation action to correct the position. This allows the Contractor to control its risk contingency effectively on an on-going basis and identify additional risks on time to escalate to the Employer.

5.7 RECOMMENDATIONS

Risk management plays an important role in the success of large-scale construction projects. The parties should review the contract carefully and allocate the risks to the party best capable to handle it and negotiate specific risks relating the current project in advance.

Smith et. al (2014) recognize the importance of undertaking risk management in the early stages of project. As explained in Figure 5.20, the earlier the risk management is conducted, the less the negative impact on the project and it is more likely that the cost of the whole project will be reduced. In contrast, the cost spent to mitigate risk will increase as the project goes on. The later the project mitigates a risk, the higher the cost.

Therefore, the Employer, the Contractor, the Consultant, and the Sub-contractor should all set up an appropriate risk management approach at the beginning of the project. Ideally, each organization establishes its own risk management procedure before entering into the contract. It would be more efficient if all parties act proactively to undertake risk management and participant with the risk mitigation discussion and action implementation.

For large construction projects, it is important to establish an appropriate risk management strategy from the beginning of the project and delegate for a competent risk manager to undertake the effective risk management throughout the project life

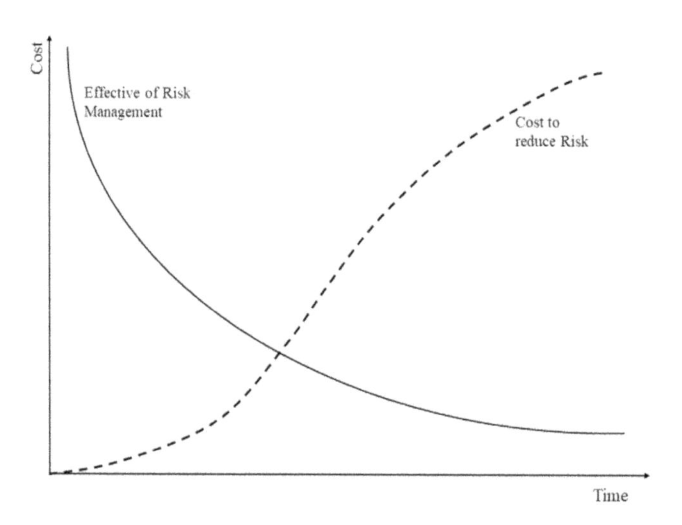

FIGURE 5.20 Risk management effectiveness in the project life cycle.

cycle. It is good practice to set up an integrated risk management system to allow all participants to contribute and engage in the on-going risk management process. In the UK, the most well-used risk management database is the Active Risk Manager (ARM) by Sword Active Risk, which consolidates all risk and opportunity input from the contractors and/or the sub-contractors in a centralized database. In addition, the quantitative risk analysis can be undertaken using Oracle Primavera Risk Analysis (PRA) or other similar software, e.g., Palisade's @RISK, Deltek Acumen Risk, and Polaris.

The comprehensive quantitative risk analysis can be a good tool to analyze the overall impact and multiple factors on the project completion and the overall cost, as well as to re-check how realistic both the completion date and project outturn forecast are provided by the planner and estimator under the traditional approach. It can facilitate identification of the potential key issues within the project and early mitigation. Even if the chance to mitigate the risks is low, it can provide senior management with a realistic view of where the project is likely to end up and advanced arrangement of resource or funding can be prepared.

Furthermore, both the project manager and risk manager should comply with the contract requirement to handle early warnings and changes and collaborate with others to mitigate potential risks at the early stage.

REFERENCES

Book/Article

Axelos, 2017. *Managing Successful Projects with PRICE2*. 2017 Edition. Norfolk, UK: TSO (The Stationery Office) / Williams Lea Tag.

HM Treasury UK. 2004. *The Orange Book, Management of Risk -Principles and Concepts.*

ISO. 2018. *ISO31000 Risk Management - Guidelines.* Geneva, Switzerland: International Standard Organization.

PMI. 2015. *Pulse of the Profession: Capturing the Value of Project Management.* https://www.pmi.org/-/media/pmi/documents/public/pdf/learning/thought-leadership/pulse/pulse-of-the-profession-2015.pdf, Accessed on 26 October 2019.

PMI. 2017. *A Guide to the Project Management Body of Knowledge, PMBOK Guide.* 6th Edition. Newtown Square, PA: Project Management Institute.

PMI. 2019. *The Standard for Risk Management in Portfolios, Programs, and Projects.* Newtown Square, PA: Project Management Institute.

Smith, N. J., Merna, T. and Jobling, P. 2014. *Managing Risk in Construction Projects.* 3rd Edition. West Sussex, UK: John Wiley & Sons Incorporated.

Contract

FIDIC. 1999. *Conditions of Contract for Construction.* 1st Edition. (1999 Red Book). Geneva, Switzerland: The Fédération Internationale des Ingénieurs-Conseils.

FIDIC. 2017. *Conditions of Contract for Construction.* 2nd Edition. (2017 Red Book). Geneva, Switzerland: The Fédération Internationale des Ingénieurs-Conseils.

FIDIC. 2017. *Conditions of Contract for EPC / Turnkey Project.* 2nd Edition. (2017 Yellow Book). Geneva, Switzerland: The Fédération Internationale des Ingénieurs-Conseils.

FIDIC. 2017. *Conditions of Contract for Plant & Design Build.* 2nd Edition. (2017 Silver Book). Geneva, Switzerland: The Fédération Internationale des Ingénieurs-Conseils.

NEC. 2013. *NEC3 Engineering and Construction Contract.* London, UK: Thomas Telford Ltd.

NEC. 2017. *NEC4 Engineering and Construction Contract.* London, UK: Thomas Telford Ltd.

6 Change Management

6.1 INTRODUCTION

Due to the complex nature of large construction projects, change is inevitable. Contract provisions in standard forms of contract usually provide the Contractor with the rights to claim or compensate changes under the contract. The contract usually sets out the situations in which the Contractor is entitled to apply for a change as well as the required procedure to execute the change. This chapter explains how change management has been undertaken under FIDIC, NEC, and JCT forms of contract and how the provisions provided in the contract are related to change management principles under PMBOK and PRINCE2. It then recommends the effective change management procedure integrated with both contract requirements and standard project management principles.

6.2 VARIATIONS

Due to the complex nature of large construction projects, the Employer needs to serve a scope in certain detail to allow the Contractor to price the work accurately in its bid. However, the more detail there is in the scope document, the more changes that are likely to occur during the actual implementation. Furthermore, some contracts may be formed before the scope of work has been fully finalized.

Once the contract is incorporated, the Contractor is obliged to complete the works in accordance with the scope of work set out in the contract. In the absence of the variation clauses, the Employer cannot change the contract, hence the Contractor cannot undertake additional work.

In order to avoid changing the contract every time the Employer would like to make a change to the work, most standard forms of contract contains clauses to allow the Employer to make changes via variation instruction. The contract generally illustrates a list of circumstances which constitute a variation.

The Contractor generally needs to obey the Employer's instructions and will be compensated with additional time and money associated with the instruction. The contract generally also provides the procedure to assess the variations. On the one hand, the variation clauses provide flexibility to the Employer to make changes to the original scope if it is no longer appropriate; on the other hand, the Contractor receives compensation to incorporate such changes. Such flexibility reduces the risk for the fixed scope of work at the beginning of the project and also reduces disputes between the contract parties.

Variation may arise due to adding to, reducing, or substituting the work, change of conditions, or restriction to the Contractor. However, if the Employer instructs the Contractor to omit certain work and then gives it to another party, that would constitute breach of contract. For example, in *Amec Building Company Limited v Cadmus Investment Company Limited (1996) 51 ConLR 105*, the Employer omitted the work and gave it to another Contractor. The Court held that the Employer cannot withdraw the original contract by omitting the work, and awarded the Contractor compensation. The Contractor can request for changes or variations when he think he is required to carry out works outside the contract or the change of the Contractor assumed conditions prevent him from carrying out works as he originally planned for. The contract may also incentivize the Contractor to propose positive changes that may lead to savings of the overall project, e.g., value engineering. In such situations, the Employer and the Contractor will share the savings, which is a win-win solution for both parties.

6.3 CLAIM

Apart from the variations, which are usually initiated by the instruction from the Engineer under FIDIC, from the *Project Manager* under NEC, or the Employer under JCT, another component that may lead to changes to the project is the Contractor's and/or the Employer's claim. In general terms, a claim refers to a contract party obtaining an entitlement from the other party for an event which occurred during the project implementation. Both the Contractor and the Employer have the right to claim in construction projects. The typical claim for a Contractor includes extension of time or time at large, loss and expense or quantum maturity, and variations in scope change. The Employer's claim often relates to liquidated damages, defects, and professional duty of care. Based on the sources that give rise to a claim, claims can be classified as claim under the contract, claim of breach of contract, and claim outside the contract.

6.3.1 Claim Inside the Contract

Claims under the contract generally refer to the contract provisions which provide a party with the rights to claim compensation when pre-agreed events arise. The Contractor can claim extension of time and associated cost for an Employer's risk event in the typical construction contract. The Employer can claim liquidated damages if the Contractor fails to complete the work by contract completion date. The provisions give rise to the Contractor making a claim not only under the express provisions, but also the implied terms from the statutory and/or common law. If a party breaches either the express terms or implied terms of the contract, the other party can claim breach of contract.

6.3.1.1 Express Terms

Standard forms of contract often includes provisions to provide the Contractor with the rights to claim for additional time to avoid the prevention principle as in *Peak Construction Ltd v McKinney Foundations Ltd* [1970] 1 BLR 111 and for time to be set at large as in *Multiplex Constructions (UK) Ltd v Honeywell Control Systems Ltd* EWHC 447 (TCC) as discussed in Chapter 3. Such situations also provide the Contractor with the right to claim for additional cost to avoid payment under *quantum meruit* as discussed in Chapter 4.

There are three types of Contractor's claim under the contract provisions, namely claims for extension of time, claims for additional cost, or claim for both additional time and cost. These claims are usually provided with the entitlement under the contract such as variations in general term, Relevant Events under JCT, compensation events under NEC, and Engineer's instruction of variation and/or Contractor's claims under FIDIC. Generally, the Contractor needs to comply with the notice requirements stipulated in the contract to fulfil a successful claim. The contract usually also provides an evaluation method to assess the variation amount. In general, there are three approaches to evaluate the variation: 1) based on the quantities completed, 2) fair allocation of change of quantities, and 3) fair assessment of additional work.

A successful claim for extension of time does not automatically entitle the winning party to claim for additional costs. Some contracts expressly provide the direct link between time and cost such as NEC, which provides both time and cost entitlement for all compensation events; whereas other contracts usually delink the time claim and money claim, such as JCT and FIDIC. The main reason to separate the time claim from the money claim is the consideration of neutral events which are not the fault of neither party. Therefore, the neutral event, such as adverse weather conditions, generally only give the Contractor the right to claim additional time, but not additional cost. If the Contractor wants to claim for additional costs, it needs to prove causation through the "But for" test laid out in *Barnett v Chelsea & Kensington Hospital* [1969] 1 QB 428 or the balance of probabilities as seen in *Walter Lilly & Company Ltd v Mackay and another* [2012] EWHC 1773 (TCC).

Furthermore, an instruction issued by the Employer's representative such as the Contract Administrator under JCT, the Engineer under FIDIC, or the *Project Manager* under NEC, gives rise to a variation for the Contractor to claim for additional payment and/or time.

6.3.1.2 Implied Terms

The implied terms can be established through the common law and statutes in the jurisdiction governing the contract. In practice, the implied terms under SGA 1979 and SGSA 1982 are often used to establish the Employer's claim in construction dispute cases.

The implied terms of satisfaction of quality come under both Section 14(2) Sales of Goods Act (SGA) 1979 and Section 4(2) Supply of Goods and Services Act (SGSA) 1982.

S14(2) of SAG 1979 provides:

> *Where the seller sells goods in the course of a business, there is an implied term that the goods supplied under the contract are of satisfactory quality.*
>
> (2A) *For the purposes of this Act, goods are of satisfactory quality if they meet the standard that a reasonable person would regard as satisfactory, taking account of any description of the goods, the price (if relevant) and all the other relevant circumstances.*
>
> (2B) *For the purposes of this Act, the quality of goods includes their state and condition and the following (among others) are in appropriate cases aspects of the quality of goods—*
>
> (a) *fitness for all the purposes for which goods of the kind in question are commonly supplied,*
>
> (b) *appearance and finish,*
>
> (c) *freedom from minor defects,*

> (d) *safety, and*
> (e) *durability.*
>
> (2C) *The term implied by subsection (2) above does not extend to any matter making the quality of goods unsatisfactory—*
> (a) *which is specifically drawn to the buyer's attention before the contract is made,*
> (b) *where the buyer examines the goods before the contract is made, which that examination ought to reveal, or*
> (c) *in the case of a contract for sale by sample, which would have been apparent on a reasonable examination of the sample.*

S4(2) of SGSA 1982 provides:

> *Where, under such a contract, the transferor transfers the property in goods in the course of a business, there is an implied condition that the goods supplied under the contract are of satisfactory quality.*

The obligation of fitness for purpose may also be implied to the Contractor under Section 14(3) Sales of Goods Act (SGA) 1979 and Section 4(4) & 4(5) of the Supply of Goods and Services Act (SGSA) 1982.

S14(3) of SAG 1979 provides:

> *Where the seller sells goods in the course of a business and the buyer, expressly or by implication, makes known—*
>
> (a) *to the seller, or*
> (b) *where the purchase price or part of it is payable by instalments and the goods were previously sold by a credit-broker to the seller, to that credit-broker,*
>
> *any particular purpose for which the goods are being bought, there is an implied [F1term] that the goods supplied under the contract are reasonably fit for that purpose, whether or not that is a purpose for which such goods are commonly supplied, except where the circumstances show that the buyer does not rely, or that it is unreasonable for him to rely, on the skill or judgment of the seller or credit-broker.*

S4(4–5) of SGSA 1982 provides:

> *(4) Subsection (5) below applies where, under a relevant contract for the transfer of goods, the transferor transfers the property in goods in the course of a business and the transferee, expressly or by implication, makes known -*
>
> (a) *to the transferor, or*
> (b) *where the consideration or part of the consideration for the transfer is a sum payable by instalments and the goods were previously sold by a credit-broker to the transferor, to that credit-broker, any particular purpose for which the goods are being acquired.*
>
> *(5) In that case there is (subject to subsection (6) below) an implied condition that the goods supplied under the contract are reasonably fit for that purpose, whether or not that is a purpose for which such goods are commonly supplied.*

The consultant may also be subject to the implied terms of providing service with reasonable skill and care under Section 13 of the Supply of Goods and Services Act 1982 and Section 29 of the Consumer Rights Act 2015.

Section 13 of SGSA 1982 provides implied terms about care and skill, as follows:

In a [relevant contract for the supply of a service] where the supplier is acting in the course of a business, there is an implied term that the supplier will carry out the service with reasonable care and skill.

Section 29 of the Consumer Rights Act 2015 provides Service to be performed with reasonable care and skill:

(1) Every contract to supply a service is to be treated as including a term that the trader must perform the service with reasonable care and skill.
(2) See section 54 for a consumer's rights if the trader is in breach of a term that this section requires to be treated as included in a contract.

For example, in the case of *Eckersley v. Binnie & Partners* (1988) 18 Con LR 1, the Court of Appeal held that the Engineer providing the professional service had average skill and care obligation.

6.3.2 Claim Outside the Contract

Claim outside the contract generally relies on the applicable law governing the contract, e.g., claim under tort, misrepresentation, frustration, and relevant statues. For example, under English law, the parties can make a claim under law of tort, e.g., duty of care. Other Acts relevant to construction works can also give rise to a party's claim. For example, the Housing Grants, Construction and Regeneration Act (HGCRA) 1996.

In addition, there is a presumption that the contracting party does not intend to abandon any remedies for breach of contract arising from the operation of law as in *Modern Engineering Ltd v Gilbert Ash Ltd* [1974] A.C. 689.

This chapter focuses on the claim under the contract provisions. The claim for breach of contract and claim outside the contract is not within the scope of this chapter.

6.4 CHANGE UNDER FIDIC

FIDIC 1999 suite of contract provides the Employer's claim in Clause 2.5 and the Contractor's claim in Clause 20. In FIDIC 2017 suite, Clause 20 combined both the Employer's claim and the Contractor's claim. The FIDIC contract requires dealing with the change as it arises. If the Contractor demands additional time and/or cost for undertaking certain variation works outside the contract, it needs to serve notices to the Engineer within a certain time limit. Clause 20 of FIDIC forms sets out the procedure for both the Contractor's claim and the Employer's claim.

Like NEC, FIDIC 2017 introduces sanction clauses to incentivize both contract Parties as well as the Engineer to perform within the time limit set out in the contract. Both the Contractor and the Employer need to submit a notice to the Engineer within 28 days of their awareness of a claim. In the case of failure to notify the Engineer with 28 days, the claim will be time barred if the Engineer gives the claiming Party notice of the Engineer's failure within 14 days of Notice of Claim. Following submitting the Notice of Claim, the claiming Party also needs to provide detailed claim documents within 84 days (extended from 42 days in the 1999 suite) of the event arising. Failure to provide the claiming details within 84 days will result in the Notice of Claim being deemed lapsed, if the Engineer notifies the claiming Party of his failure within 14 days of the time limit expiring. The claim procedure under FIDIC 2017 Red Book will be discussed further in Section 6.4.3.

6.4.1 Contractor's Claim

Under FIDIC 2017, if the Contractor considers he is entitled to an extension of time, then he must serve two notices as follows:

1) Sub-Clause 8.5 identifying the entitlement cause/s
2) Sub-Clause 20.2.1 Notice of Claim

Whereas, for the money claim, the Contractor only needs to serve one notice under Sub-Clause 20.2.

Clause 8.5 provides express provisions for the causes of delay as follows (FIDIC, 2017):

a) variation
b) a reason provided in the Contract
c) adverse climate conditions
d) unforeseen shortages of personnel or goods
e) prevention by the Employer

The variation under Sub-Clause 8.5(a) is then subject to agreement of time and cost under Clause 13 "Variation and Adjustment." Table 6.1 lists out the other causes of delay under FIDIC 2017 that give an entitlement to extension of time. This constitutes the causes in Clause 8.5(b).

The above entitlement to extension of time is then subject to Sub-Clause 20.2 for a valid claim. In particular, under Sub-Clause 20.2.1, the Contractor is obliged to submit a "Notice of Claim" of an entitlement to additional time or money to the Engineer as soon as practicable but no later than 28 days after his awareness of the Relevant Event. In the case of failure to notify the Engineer with 28 days, the Contractor's entitlement to claim any time or money will be time barred if the Engineer gives the claiming party notice of his failure within 14 days of "Notice of Claim". For example, in *Obrascon Huarte Lain SA v Her Majesty's Attorney General for Gibraltar [2014] EWHC 1028 (TCC)*, although the Contractor had solid ground to claim extension of time, the Employer successfully defended the claim as the Contractor failed to comply with the time limit in accordance with the contract. The Contractor is also required to provide detailed information to substantiate the claim.

Figure 6.1 demonstrates how the relationship of the clauses give rise to entitlement of extension of time under FIDIC 2017.

All entitlements to time claims are also applicable to cost claims, except for Sub-Clause 8.6, which can only be used for time claims and not cost claims.

TABLE 6.1
Additional Clauses Give Rise to Claim Extension of Time under Clause 8.5(b)

Clause	Description
1.9	Delayed Drawings or Instructions
1.13	Delay or failure to obtain permits
2.1	Right of Access to the Site
4.6	Unforeseeable request to co-operate
4.7	Setting Out
4.12	Unforeseeable Physical Conditions
4.15	Changes to access route by Employer or third parties
4.23	Archaeological and Geological Findings
7.4	Contractor's Testing delayed by Employer
7.6	Remedial work attributable to the Employer
8.6	Delays Caused by Authorities
8.10	Consequences of Employer's Suspension
10.2	Remedial work attributable to the Employer
10.3	Interference with Tests on Completion
11.7	Delay to access for rectification during Defects Notification Period
11.8	Search for cause of a defect which is not Contractor's responsibility
13.6	Adjustments for Changes in Laws
16.1	Suspension by Contractor
16.2.2	Delay of notice period for termination
16.3	Work instructed after termination
17.2	Loss/damage to the Works due to Employer's risk event
18.4	Consequences of an Exceptional Event

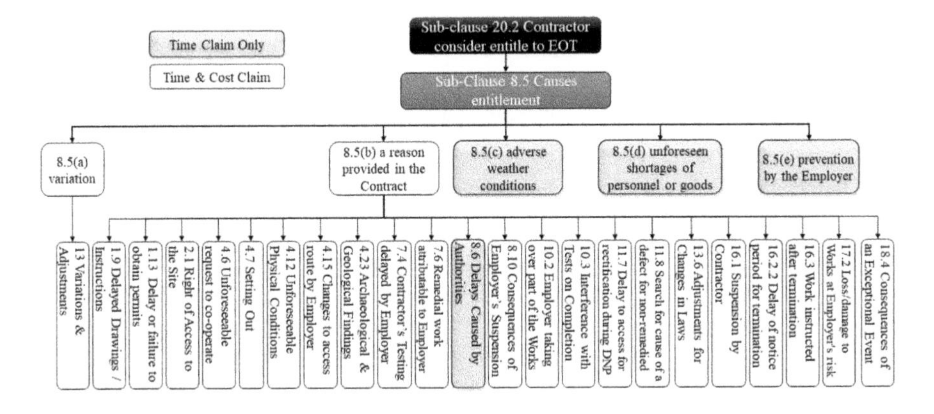

FIGURE 6.1 Clause structure for claims under FIDIC 2017 Red Book.

TABLE 6.2
Additional Clauses Give Rise to Claim Cost under Clause 8.5(b) in FIDIC 2017 Red Book

Clause	Description	With Profit?
1.9	Delayed Drawings or Instructions	Yes
1.13	Delay or failure to obtain permits	Yes
2.1	Right of Access to the Site	Yes
4.6	Unforeseeable request to co-operate	Yes
4.7	Setting Out	Yes
4.12	Unforeseeable Physical Conditions	No
4.15	Changes to access route by Employer or third parties	No
4.23	Archaeological and Geological Findings	No
7.4	Contractor's Testing delayed by Employer	Yes
7.6	Remedial work attributable to the Employer	Yes
8.10	Consequences of Employer's Suspension	No
10.2	Remedial work attributable to the Employer	Yes
10.3	Interference with Tests on Completion	Yes
11.7	Delay to access for rectification during Defects Notification Period	Yes
11.8	Search for cause of a defect which is not Contractor's responsibility	Yes
13.6	Adjustments for Changes in Laws	No
16.1	Suspension by Contractor	Yes
16.2.2	Delay of notice period for termination	Yes
16.3	Work instructed after termination	Yes
17.2	Loss/damage to the Works due to Employer's risk event	Yes
18.4	Consequences of an Exceptional Event	No

6.4.1.1 Cost Claims

Similar to time claims, Table 6.2 provides the causes that the Contractor is entitled to claim additional cost under Sub-Clause 8.5(b) of the FIDIC 2017 Red Book. Under some circumstances, the Contractor is also entitled to claim profit in the cost claim.

As shown in Table 6.2, the elements due to the Employer's wrongdoing would give the Contractor entitlement to claim profit, for example, delay in drawings or instructions, giving access of site, setting out, and Contractor suspension due to Employer's fault.

6.4.2 Employer's Claim

Table 6.3 lists the clauses that give rise to the Employer claiming damages against the Contractor.

In contrast to the Contractor's claim, the Employer's claim may allow the Employer to deduct the contract price directly and bring the Time for Completion forward.

6.4.3 Claim Process

FIDIC 2017 made a substantial change to the procedure of claim, including deemed acceptance for "Notice of Claim" under Sub-Clause 20.2.2, deemed lapse for

TABLE 6.3
Clauses Entitle the Employer's Claim under FIDIC 2017 Red Book

Clause 20.2.1 required notification	
7.5	Defects and Rejection
7.6	Remedial Work
8.7	Rate of Progress
11.3	Extension of Defects Notification Period
15.4	Payment after Termination for Contractor's Default
No notification required	
4.19	Temporary Utilities
4.2	Performance Security
Clauses allow the Employer claim and deduction	
4.2	Performance Security
5.2	Nominated Sub-contractors
9.2	Delayed Tests
10.2	Taking Over Parts
11.4	Failure to Remedy Defects
11.6	Further Tests after Remedying Defects
11.11	Clearance of Site
19.1	General Requirements
19.2	Insurance to be Provided by the Contractor

"Notice of Claim" under Clause 20.2.4, and deemed rejection of the detailed claim under Sub-Clause 3.7.3(i). Unlike the NEC contract, while the deemed acceptance is the complete end of the process, FIDIC provides the opportunity for the other Party to submit its disagreement of the deemed acceptance. For the deemed rejection of the claim, the Contractor can only refer the dispute to the DAAB. Figure 6.2 demonstrates the claim procedure under FIDIC 2017 Red and Yellow Books.

Another new feature of FIDIC 2017 is providing "Claims of continuing effect" under Sub-Clause 20.2.6.

6.5 COMPENSATION EVENT UNDER NEC

The change management under the NEC contract is implemented through the compensation event (CE). The philosophy of the compensation event under the NEC contract is that if the event does not arise due to the *Contractor*'s fault, the *Contractor* is entitled to be compensated for the effect on the Prices, Key Dates, and Completion Date. Core Clause 6 sets out the main entitlement of the compensation event and the procedure that the *Project Manager* and the *Contractor* need to follow.

The NEC contract sets out a clear process to drive the implementation of the compensation event and sets out sanctions for the parties that fail to comply with the process. The compensation event has a condition precedent nature, and failure to notify the compensation event within the eight-week limit can waive the *Contractor*'s rights to claim the consequences.

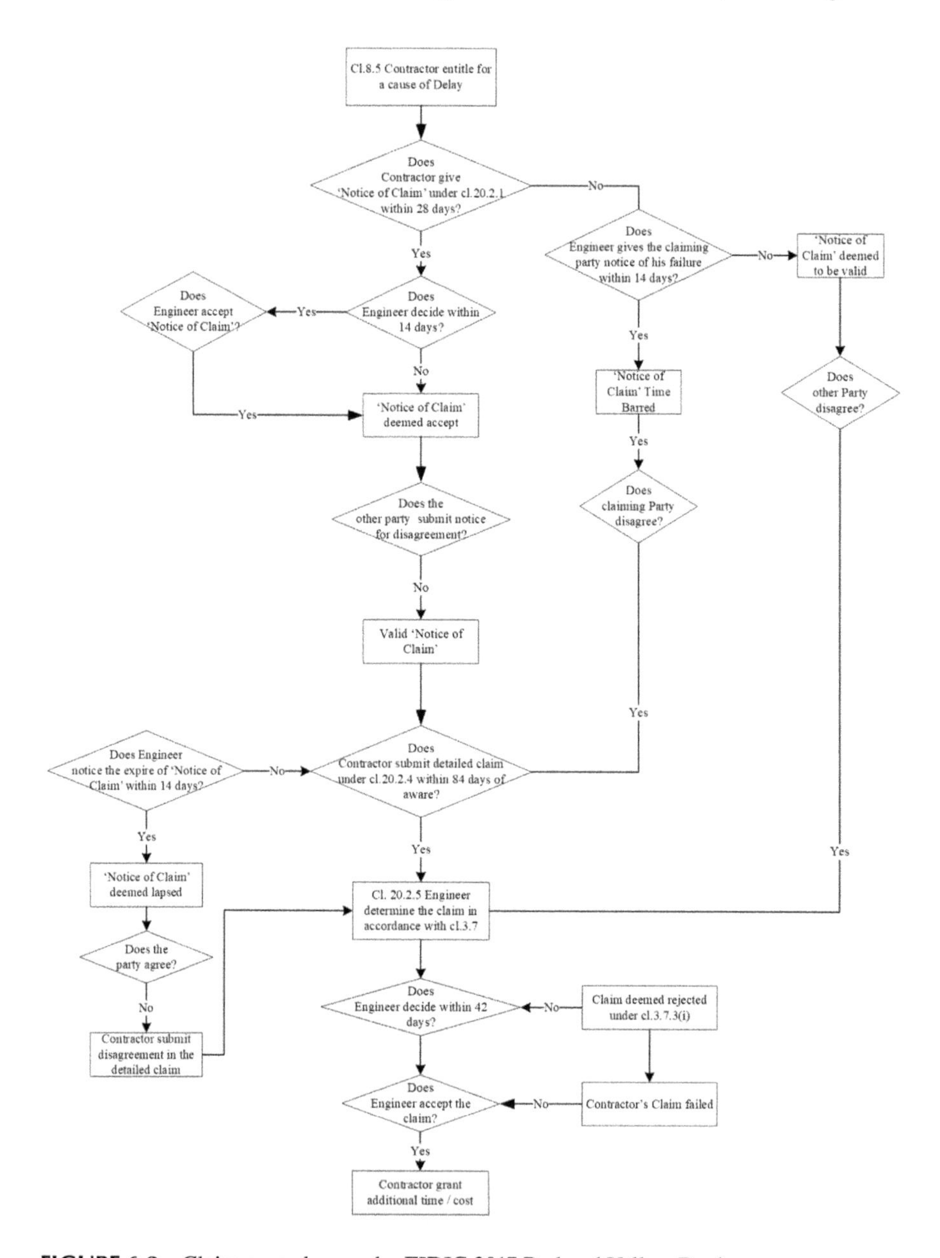

FIGURE 6.2 Claim procedure under FIDIC 2017 Red and Yellow Books.

NEC deals with the effects of time and cost together and it deals with those in Clauses 60 to 65. Clause 60.1 set out 21 entitlements for the *Contractor* claim compensations. The list can also be extended through Clause 60.1(15) to additional the *Client*'s liability under the third bullet of Clause 81, which can be defined in the Contract Data part one.

Similar to Sub-Clause 20.1 under FIDIC 2017, Clause 61.3 "notifying compensation events" are also condition precedent. For example, in *Multiplex v Honeywell*, although

Honeywell was entitled to £11 million of compensation works, it was liable to serve the notice within eight weeks as required under Clause 61.3 of the NEC3 contract. Consequently, the judge held that failure to serve notification of the compensation event amounted to breach of contract and the *Contractor* waived its right to claim.

Following the *Project Manager*'s acceptance of the notification of the compensation event under Clause 61.4, the Contractor is required to provide a quotation within three weeks under Clause 62.2, considering both time and cost, plus associated risk allowance in accordance with Clause 63.8. The *Project Manager* then needs to reply to the *Contractor*'s quotation within two weeks, unless the duration is extended following agreement by both parties under Clause 62.5.

Under Clause 65, the *Project Manager* can request alternative quotations to deal with the change in terms of different time and/or cost options. This allows the *Project Manager* to select the quotation that best suits the *Client*'s requirements. In addition, the *Project Manager* can better control the remaining time and cost and provide the *Client* with a better forecast of the Completion Date and Total of the Price. This enables actions to be taken in a timely manner where budget/time constraints are crucial.

Following receiving the *Contractor*'s quotation, the *Project Manager* will assess the quotation and reply within two weeks. Once the *Project Manager* responds with their decision, in accordance with Clause 66.1, the compensation event is implemented. Clause 66.2 further provides that once implementation, no retrospective review can be undertaken to the price or time of the compensation event. If the *Contractor* believes the *Project Manager* underestimated the cost or time of the compensation event, the only way to resolve this is through the dispute resolution provided under secondary option Clause W1 contractual adjudication, W2 statutory adjudication, or W3 Dispute Avoidance Board. If the dispute cannot be resolved under adjudication or dispute board, then the *Contractor* can refer the case to the tribunal, which leads to the local court or arbitration. In such circumstances, the *Contractor* has to check with the relevant contract provisions.

6.5.1 Entitlement

There are four sources of compensation event under the NEC contract, including:

1) 21 reasons in NEC4 (NEC, 2017) or 19 reasons in NEC3 (NEC, 2013) set out under Clause 60.1
2) three additional reasons in relation to change of quantity provided in Clauses 60.4, 60.5 and 60.6 for Option B and D contract
3) Secondary Options, including:
    - X2.1 changes of law
    - X12.3(6)–(7) change of Patterning information or programme
    - X14.2 Delay in making advanced payment
    - X15.2 *Contractor*'s corrects a Defect for which it is not liable for
    - X21.5 Whole life cost (new in NEC4)
    - Y2.5 Suspension of performance
4) Potential entitlement under Option Z clauses

6.5.1.1 Events Resulting from the *Client* or Their Representative

Clause 60.1 sets out 21 main reasons for the compensation event. The following elements are caused by the *Client* and/or their representative:

- 60.1(1): the *Project Manager*'s instruction to changes, in particular change of the Scope of work
- 60.1(2): the *Client* fails to provide access or possession of the Site by the date shown in the Accepted Programme
- 60.1(6): the *Project Manager* or the *Supervisor* fails to reply to a communication within the time limited required in the contract
- 60.1(8) the *Project Manager* or the *Supervisor* changes a previous decision
- 60.1(9) the *Project Manager* withhold acceptance for a reason not stated in the contract
- 60.1(10) the *Supervisor* instructs the *Contractor* to search for Defect, but no Defect is found
- 60.1(15), the *Client* take over the works earlier than the contract stated date
- 60.1(18) the *Client* breaches of contract but not a compensation event

Clause 13.4 requires the *Project Manager* to communicate to the *Contractor* for his submission and provide a reason for non-acceptance. Clause 13.8 further provides that the *Project Manager* withhold acceptance of the *Contractor*'s submission for a reason stated in the contract that does not constitute a compensation event. In order to understand what constitutes the entitlement under Clause 60.1(9), the *Project Manager* withholds acceptance for a reason not stated in the contract; it is important to understand what the non-acceptance reasons are already provided in the contract. Clauses require the *Contractor*'s submission for the *Project Manager*'s acceptance under NEC4 are listed as follows (NEC, 2017):

- 16.1 the *Contractor* may submit a proposal to change the Scope and the *Project Manager* needs to respond within four weeks and provide reasons for non-acceptance
- 16.2 the *Contractor* may submit a proposal to change the Working Areas. The reasons for the project non-acceptance are:
    1) the proposed area is not necessary for Providing the Works
    2) it is used for work outside the contract
- 21.2 the *Contractor* submits its design to the *Project Manager* for acceptance; the reasons for non-acceptance is that the design does not comply with:
    1) the Scope, or
    2) the applicable law
- 23.1 the *Contractor* submits their design of Equipment to the *Project Manager* for acceptance; the reasons for non-acceptance is that the *Contractor*'s design of the Equipment does not allow the *Contractor* to Provide the Works in accordance with:
    1) the Scope
    2) the Project Manager accepted the *Contractor*'s design, or
    3) the applicable law

- 24.1 the *Contractor* submits the name, relevant qualification, and experience of a replacement person to the *Project Manager* for acceptance; the reason for non-acceptance is that the proposed personnel is not as good as the existing personnel in terms of qualifications and experience.
- 26.2 the *Contractor* submits the proposed Sub-contractor to the *Project Manager* for acceptance; the reason for non-acceptance is that the proposed Sub-contractor does not "*allow the Contractor to Provide the Works.*"
- 31.3 the *Contractor* submits the programme to the *Project Manager* for acceptance. The *Project Manager* needs to response within two weeks, and the reasons for non-acceptance are:
    1) the *Contractor*'s programme is not practicable
    2) the programme does not show the information required under Clause 31.2
    3) the programme is not realistic, or
    4) the programme does not comply with the Scope
- 84.1 the *Contractor* submits the contract required insurance certification to the *Project Manager* for acceptance before the *start date*, and the reasons for non-acceptance are:
    1) the insurance does not comply with the contract
    2) the insurer's commercial position is not sufficient for the insured liabilities
- X13.1 the *Contractor* submits a performance bond to the *Project Manager* for acceptance, and the reason for non-acceptance is that the bank or insurer's commercial position is not sufficient for the insured liabilities.
- X14.2 the *Contractor* submits the advanced payment bond to the *Project Manager* for acceptance, and the reason for non-acceptance is that the bank or insurer's commercial position is not sufficient for the insured liabilities.
- Y1.4 the *Contractor* submits the banking arrangements for the Project Bank Account to the *Project Manager* for acceptance, and the reason for non-acceptance is that the bank does not provide payment to be made in accordance with the contract.
- Y1.6 the *Contractor* submits the proposed Supplier to the *Project Manager* for acceptance, and the reason for non-acceptance is that "*the addition of the Supplier does not comply with the Scope.*"

The *Project Manager*'s non-acceptance for any reason other than from the above list would constitute a compensation event under Clause 60.1(9).

6.5.1.2 Events Outside the *Client*'s Control

In addition, the following entitlements arise outside the *Client*'s control, but the *Client* bears the risk:

- 60.1(7) objects found within the site, e.g., archeology, fossils
- 60.1(12) unforeseen physical conditions
- 60.1(13) adverse weather conditions

- 60.1(14) the *Client*'s liability specified in the Contract Data part one, and under Clause 80.1
- 60.1(19) an event prevents the *Contractor* completing the works, such as the *force majeure* event under FIDIC 1999 edition

Under Options B and D contracts relating to Bills of Quantities, there are three additional reasons that give rise to the compensation event. Clause 60.4 provides difference in quantity affecting unit cost, Clause 60.5 provides difference in quantity delays Completion or Key Date, and Clause 60.6 provides correcting a mistake in the bill of quantities.

In addition, Table 6.4 lists the relevant clauses and the reasons that give for entitlement to a compensation event under the secondary options.

NEC4 provides a new clause Option X21 "Whole Life Cost" to give the *Contractor* the opportunity to propose changes, which can reduce the project whole life cost and not only includes the project implementation cost but is more related to cost of operation and maintenance post Completion.

If the *Project Manager* is willing to consider the change, the *Contractor* needs to submit a detailed quotation with the detailed proposal, the forecast of total cost reduction, as well as the cost to undertake the changes, and the revised programme to show the impact on the Completion Date and/or any relevant Key Dates.

The *Project Manager* then needs to review the *Contractor*'s quotation and respond within the *period for reply*. If the *Project Manager* accepts the *Contractor*'s quotation, the change will be implemented and the Scope, the Price, the Completion Date, and the Key Date will be updated accordingly. However, the changes undertaken under Option X21 do not constitute a compensation event.

In addition, more bespoke entitlements for compensation events can be defined in the Contract Data part one by the *Client* as referred to Clause 60.1.(21).

6.5.2 Process

The NEC contract requests both the *Project Manager* and the *Contractor* to notify the other of a compensation event as soon as they are aware. After the *Project Manager* has accepted the compensation event notification, he then instruct the *Contractor* to

TABLE 6.4
Entitlement of Compensation Event under the Secondary Options

Clause	Entitlement
X2.1	Changes in the law after the Contract Date
X12.3(6)	Changes to the Partnering Information
X12.3(7)	Changes to the timetable for Core Group contributions of the Partners
X14.2	Delay in making an advanced payment
X15.2	*Contractor* corrects a defect for which he was not liable under their contract
X21.3	*Contractor*'s proposed scope change in order to reduce operation and maintenance cost
Y2.5	*Contractor* exercises his right under the Construction Act to suspend performance
Z	Additional conditions of contract if relevant

TABLE 6.5
Entitlement of Compensation Event under Core Clause 60.1

Contractor Notify		*Project Manager* Notify	
60.1(2)	No Access to and/or use of each part of the Site	60.1(1)	Changes to the Scope
60.1(3)	Provision by the *Client*	60.1(4)	Stop work or change a Key Date
60.1(5)	Work by the *Client* or Others	60.1(7)	Objects found within Site
60.1(6)	Late reply to a communication	60.1(8)	Change previous decision
60.1(9)	Withhold an acceptance	60.1(10)	Search but no defect
60.1(11)	Test/inspection causes a Delay	60.1(15)	*Client*'s early take over
60.1(12)	Unexpected physical conditions	60.1(17)	Correct an Assumption
60.1(13)	Adverse weather	60.1(20)	Proposed instruction is not accepted
60.1(14)	*Client*'s liability events		
60.1(16)	Materials, facilities for tests		
60.1(18)	*Client*'s breach of contract		
60.1(19)	Prevention		
60.1(21)	Additional compensation events stated in Contract Data part one		

submit a quotation for the compensation event. In accordance with Clause 63.8, the *Contractor* should include appropriate risk allowance in their quotation as well as extension of time required. Unlike other construction contracts, e.g., the FIDIC and JCT contracts, which separate the compensable time and cost entitlement based on whether the risks are the Employer's or the Contractor's responsibility; all compensation events under the NEC contract grant the *Contractor* the entitlement to claim both cost and time even in the case of natural events, such as Clause 60.1(19).

Table 6.5 lists the notification responsibility for the *Contractor* and the *Project Manager* respectively.

Similar to the FIDIC contract, there is a time bar for the *Contractor* to notify a compensation event to the *Project Manager.* If the *Contractor* fails to notify a compensation event that he is entitled to under the 13 reasons listed on the left side of Table 6.5, within eight weeks from initial awareness, the Contractor will waive his right to change under Clause 61.3. However, for the other eight reasons for which the *Project Manager* is responsible to give notification, there is no time bar.

Figure 6.3 demonstrates the general process for the compensation event under the NEC4 ECC contract.

Upon receiving notification of a compensation event, the *Project Manager* should decide whether it is a valid compensation event within one week. If it is, the *Project Manager* should instruct the *Contractor* to provide a quotation and the *Contractor* should then submit his quotation within three weeks. After receiving the quotation, the *Project Manager* should respond with a decision within two weeks. If the *Project Manager* accepts the *Contractor*'s quotation, the *Project Manager* then instructs the *Contractor* of the acceptance and the compensation event is implemented. It is important that NEC does not allow retrospective review of an implemented compensation event. If there is anything wrong with the value of the compensation event,

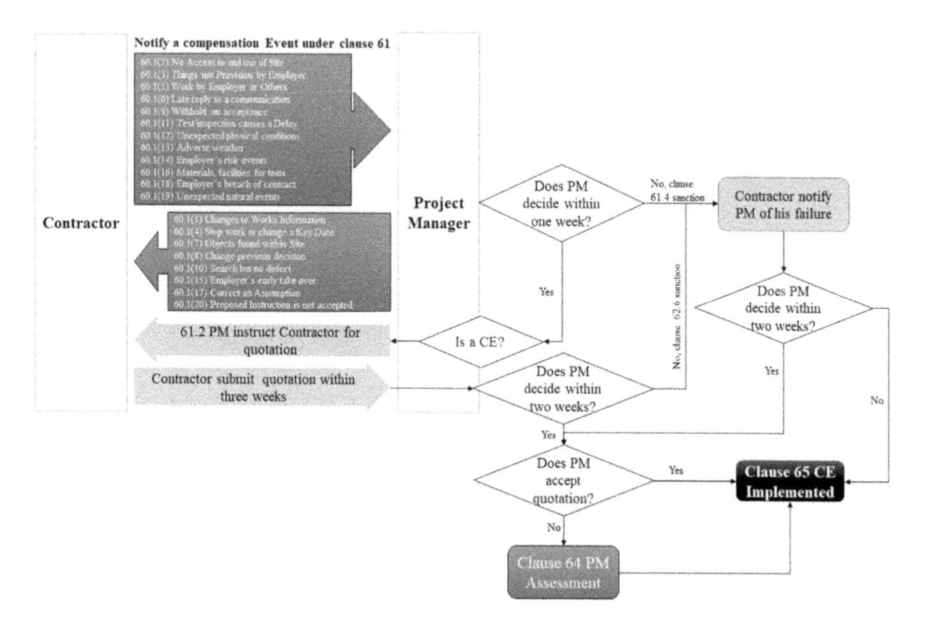

FIGURE 6.3 Compensation Event general process under NEC3 ECC contract.

the *Project Manager* can notify another compensation event if it is due to the *Project Manager*'s wrong assumption under Clause 60.1(17), otherwise the only resolution is from the adjudication provided in Options W1, W2, or W3 under the contract. It is important to notice that Clause 60.1(8) "changing a decision" does not give the *Project Manager* the right to change the value or time of the implemented compensation event, because such a right has been expressly waived in Clause 66.2.

If the *Project Manager* does not accept the *Contractor*'s quotation, he may request the *Contractor* to re-submit a quotation or undertake their own assessment under Clause 64. The NEC also sets out sanctions for the *Project Manager*'s liability. If the *Project Manager* fails to decide the *Contractor* notified compensation event within one week or the *Contractor*'s quotation within two weeks, the *Contractor* can notify the *Project Manager* of his failure under Clauses 61.4 and 62.6 respectively. If the *Project Manager* continues this failure for another two weeks, the notified compensation event or the submitted quotation is deemed to be accepted and the remaining procedure continues. However, if the *Contractor* does not notify the *Project Manager* of their failure, then the contract does not provide automatic deemed acceptance.

Another special feature of the NEC contract is that it does not require the *Contractor* to hold off on the work until the implementation of the compensation event. The *Contractor* can carry on the required works during the application of the compensation event.

Although there is a different payment mechanism for different main options of contract, e.g., Option A based on the completed activities in the Activity Schedule, and Option B based on completed works applying to fixed Bills of Quantity, the same principles for the quotation of the compensation event are applied to all main options under Clause 63.1. When preparing the quotation of a compensation event, the *Contractor* should separate the cost before and after the notification date, which is often known as

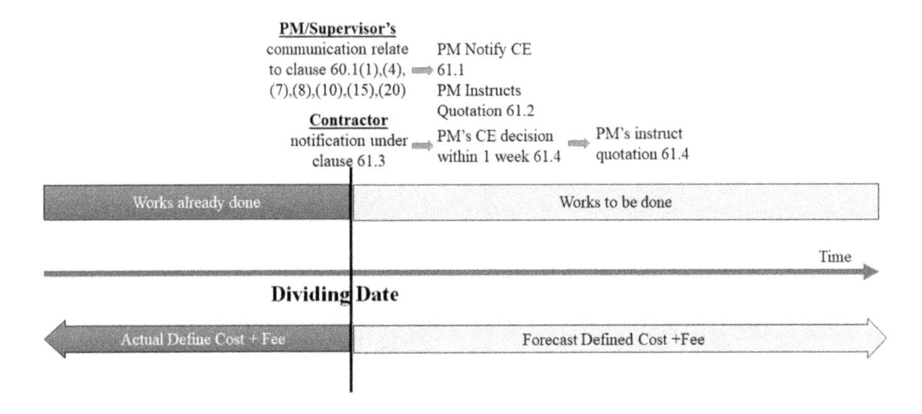

FIGURE 6.4 NEC compensation event cost build-up.

the "switch date" or "dividing date." The approaches used for calculating the cost are different before and after the switch date. As shown in Figure 6.4, before the switch date, the actual Defined Cost plus Fee is used to calculate the cost for the works already done; after the switch date, the forecast Defined Cost plus Fee is used to calculate the cost of the works to be done. The total cost for works already done and works to be done will be the total quotation of the compensation event.

Clause 62.2 requires the *Contractor* to submit "any delay to the Completion Date and Key Dates" in its quotation of the compensation event. Under the NEC contract, the contractual Completion Date can only be moved to a later date by the compensation event. The assessment of the time extended from the existing Completion Date is undertaken on the remaining activities based on the latest Accepted Programme. As discussed in Chapter 3, the *Contractor* owns the Terminal Float and the Time Risk Allowance (TRA). Consequently, the assessment of the impact compensation event should exclude these floats.

As shown in Figure 6.5, compensation event 1 dose not impact the contract Completion Date, because it uses the seven weeks total float available in the "Car Park" network path. The implementation of compensation event 1 only reduces the total float on the "Car Park" network path from seven weeks to four weeks. In contrast, compensation event 2 extends the Completion Date by two weeks, because the planned completion date has been extended for two weeks. Although there are two weeks Terminal Float between the planned completion date and the Completion Date, because the *Contractor* owned the Terminal Float, the *Client* cannot take away the Terminal Float owned by the *Contractor* during implementation of the compensation event. Therefore, in accordance with Clause 63.5, the *Contractor* is entitled to two weeks extension of the duration between the previous planned Completion and current Completion Date. The *Contractor* retains two weeks terminal float in the programme.

Clause 62.3 requires the *Contractor* to submit his quotation within three weeks and then the *Project Manager* makes his decision within two weeks. Under Clause 62.5, the *Project Manager* can extend the time for both the *Contractor*'s quotation and the *Project Manager*'s response to the quotation with the agreement of both the *Project Manager* and the *Contractor.*

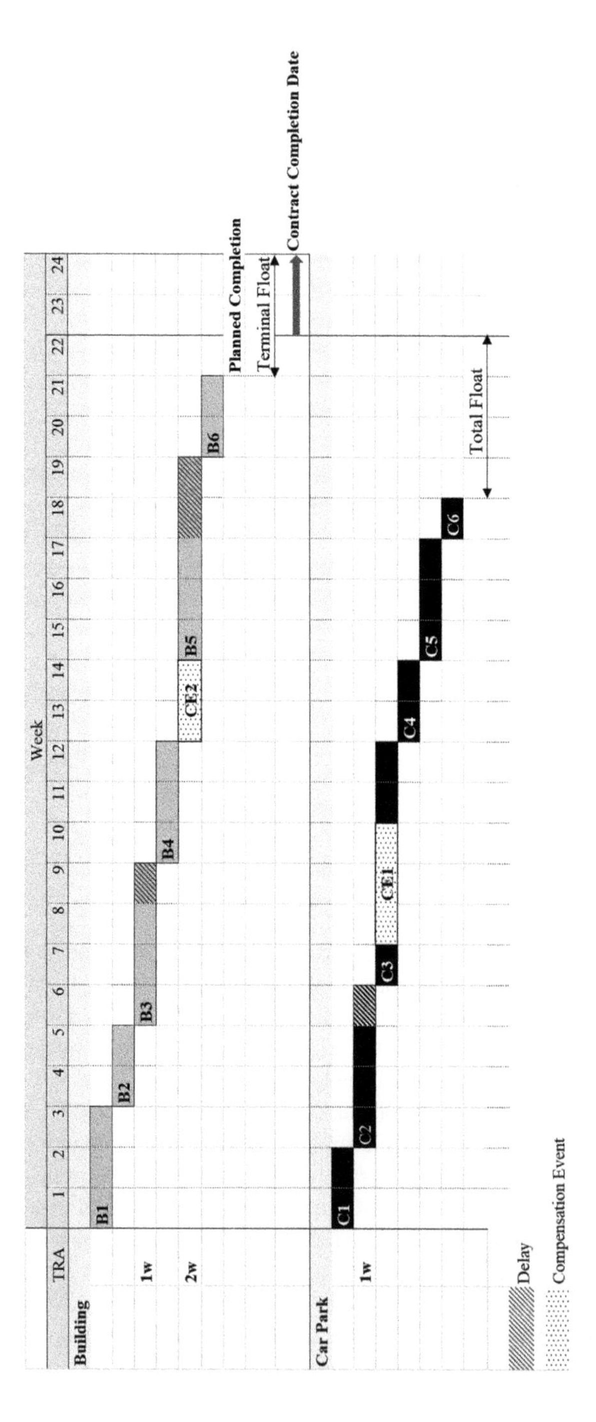

FIGURE 6.5 Compensation event time impact.

Clause 64 provides four occasions that the *Project Manager* may undertake his own assessment on the *Contractor*'s programme of a compensation event, as follows:

1) the *Contractor* does not assess the impact to the Complete Date appropriately under Clause 63.5
2) no Accepted Programme is in place
3) the *Contractor* does not submit a programme, or
4) the *Project Manager* does not accept the *Contractor*'s latest programme in accordance with Clause 31.3

If the *Project Manager* does not accept the Contractor's programme by mistake, he will be responsible for undertaking a retrospective delay analysis based on the actual delays for the extension of time in the *Project Manager*'s own assessment. If the *Project Manager* appropriately rejects the *Contractor*'s programme, then the assessment for the extension of time will be based on the forecast delay instead.

Likewise, the delay to the Key Dates is assessed based on the movement from the previous planned completion date instead of the Key Dates stated in the contract.

Therefore, the *Contractor*'s baseline programme is important for the changes of the Completion Date and Key Date in the later project implementation. As the programme network integrates with different floats in different network paths deciding whether an extension of time can be granted and how long can be granted. Both the *Contractor* and the *Project Manager* should thoroughly review the programme before submitting or approving the programme.

6.5.3 Sanctions

The NEC contract provides a specific time requirement for each party to perform his role in the process of the compensation event. There will be grounds for sanctions if a party fails to perform in accordance with the contract.

The sanction of the *Project Manager* relates to his decision of a valid compensation event and the assessment of the value of the compensation event. After the *Contractor* notifies the *Project Manager* of a compensation event under Clause 61.3, the *Project Manager* needs to reply with a decision within one week.

As discussed in Section 6.3.1, the sanction of the *Contractor* relates to the time bar for notifying of a compensation event. If the *Contractor* fails to notify the *Project Manager* of a compensation event for its entitlement under Clause 60.1(2), (3), (5), (6), (9), (11), (12), (13), (14), (16), (18), (19), and 60.1(21) if applicable, the *Contractor* loses his right to obtain compensation for such event.

In addition, if the *Contractor* does not notify of the *Project Manager*'s failure to decide a valid compensation event or reply to the quotation, the notice for compensation event or the quotation will not be treated as accepted by the *Project Manager*.

An early warning notice is not the condition precedent for the compensation event. If the *Contractor* fails to notify an early warning, which then develops into a compensation event, the *Contractor*'s right to a compensation event will not be waived.

However, the contract provides a sanction to the *Contractor*'s wrongdoings. Similar to Clause 63.5 under NEC3 (NEC, 2013), Clause 63.7 under NEC4 (NEC, 2017) states:

> *If the Project Manager has stated in the instruction to submit quotations that the Contractor did not give an early warning of the event which an experienced contractor could have given, the compensation event is assessed as f the Contractor had given the early warning.*

Timely notification of early warnings allows the project team to seek mitigation before the issues develop further and an alternative decision may be made. Thus, the intention behind such a sanction is to encourage the Parties to notify the potential issues in time in order to undertake effective risk management.

6.5.4 Changes in NEC4

Change management under the NEC contract is mainly carried out under Section 6 "Compensation Event." The NEC4 contract includes three additional reasons for a compensation event, two reasons under Clause 60.1, and one reason under Option 21. All the changes aim to provide more flexibility for the Parties.

In contrast with other construction contracts, all compensation events under the NEC contract are entitled to both time and money claims.

Under the NEC contract, there are only three circumstances that can change the contract Completion Date, as follows:

1) implementation of compensation event under Core Clause 6 to postpone the Completion Date
2) acceleration under Clause 36 to bring the Completion Date forward, and
3) change under Option 21 "whole life cost" to amend the Completion Date and/or Key Dates"

Under Clauses 16.1 and 16.2, the *Contractor* may propose a change of scope to reduce the *Client*'s payment. Clause 60.1(20) deals with the *Contractor*'s incurred cost associated with preparing a quotation for proposed instruction, which is not accepted by the *Project Manager*. This clause aims to reimburse the *Contractor*'s cost for preparing the quotation, which encourages the *Contractor* to submit the proposal if cost reduction opportunities are identified by the *Contractor*, for example through value engineering.

Likewise, Option 21 provides the *Contractor* with opportunities to propose changes to reduce the operation and maintenance cost. In order to encourage the *Contractor* to develop such proposals, the *Contractor* is entitled to compensate the cost incurred related to such a proposal.

Clause 60.1(21) provides flexibility for the contract Parties to agree additional reasons for a compensation event in accordance with specific features of the project. The additional reasons should be agreed before contract formation and the *Client* should list them out in Contract Data part one.

In addition, NEC4 also includes new Clause 65 "proposed instructions". The purpose of this clause is for the *Project Manager* to understand the potential impact on both cost and time for an intended change before the *Project Manager* issues the instruction under Clause 60.1(1).

Under Clause 65.2, the *Contractor* should submit the quotation within three weeks after receiving the *Project Manager*'s instruction and the quotation will be assessed as a compensation event under Clause 63. The *Project Manager* may provide three possible responses to the *Contractor*'s quotation as follows (NEC, 2017):

> *1) an instruction to submit a revised quotation including the reasons for doing so,*
> *2) the issue of the instruction together with a notification of the instruction as a compensation event and acceptance of the quotation or*
> *3) a notification that the quotation is not accepted.*

If the *Project Manager* does not reply within the required time limit, then the quotation is not accepted. In such circumstances, under Clause 65.3, the *Project Manager* may notify a compensation event and request the *Contractor* to submit a quotation in accordance with Clause 61.1.

6.6 VARIATIONS UNDER JCT

Like other standard forms of contract, the JCT contract includes variations or instructed change as one of the grounds for extension of time. The variation clause in JCT includes addition, omission, or any changes to the obligations of the Contractor as long as it is still within the scope of contract. The specific definition of variations in JCT can be found in reference to Clause 5.1. The JCT contract also sets out "*the imposition by the Employer of any obligation or restriction*" in Clause 5.2. The Contractor's application for payment in respect to variations is covered in Clauses 5.6–5.9; payment in respect of overheads and profit may well have been allowed for in the Contractor's variation claim.

Under JCT, the Contractor is only obliged to claim for more time and money after the risk event has occurred. In relation to extensions of time and loss and expense, the JCT contract has Relevant Matters and Relevant Events and time and money are dealt with as separate concepts.

The JCT contract does not have a programme as a contractual document. Additional time is only allowed for Relevant Events under Clause 4.2, which are listed in Clause 2.29 of the JCT Standard Building Contract (SBC) 2016 or Clause 2.26 of the JCT Design and Build (DB) 2016 contract. Loss and Expense is only paid for Relevant Matters under Clause 4.24 that materially affect the progress of the works.

The Contractor must use his "best endeavors" to prevent delay and do anything reasonable required by the Contract Administrator to proceed with the works. Under Clause 2.27, the Contractor is obliged to serve a notice of delay to the Contract Administrator in the event "*...works or any section is being or likely to be delayed....*" The Contractor need to provide further information required by the Contract Administrator which is "*reasonably necessary for the purpose of Clause 2.19.*" The Contract Administrator then assesses: 1) if completion is delayed or likely to be delayed and 2) whether the cause of the delay is a Relevant Event as listed in Clause 2.20. Then, the Contract Administrator must give a notice in writing setting out a "fair and reasonable" extension of time.

Like NEC and FIDIC, the notice is the condition precedent for the Contractor to claim under Clause 4.23. However, an application for an extension of time under Clause 2.26 is not a pre-condition to a loss and expense claim under Clause 4.23. Similarly, the Contractor also needs to serve notice for claim for money under Relevant Matters under Clause 4.24.

6.6.1 Cost Claims

As discussed in Chapter 4, the money claim under the JCT contract needs to refer to the Relevant Matters under Clause 4.21 JCT DB 2016 contract. As described in Table 6.6, the five grounds that give the Contractor the right to claim money also give it the right to claim time. The remaining nine events only allow the Contractor to claim extension of time but no additional loss and/or expenses.

6.6.2 Time Claims

As discussed in Chapter 4, the time claim under the JCT DB contract needs to refer to the Relevant Events under Clause 2.26 of JCT DB 2016 contract. Sometimes,

TABLE 6.6
Events Entitles the Contractor Claim under JCT DB 2016

No.	Event	Relevant Event	Relevant Matter
1	"Changes and any other matters or instructions under which these Conditions are to be treated as, or as requiring, a Change"	√	√
2	"Employers Instructions…"	√	√
3	"Deferment of giving of possession of the site or any Section…"	√	
4	"Compliance with Clause 3.15.1 or with the Employer's instructions under Clause 3.15.2"	√	√
5	"Suspension by the Contractor under Clause 4.11 of the performance of any or all of his obligations under this Contract"	√	
6	"Any impediment, prevention or default, whether by act or omission, by the Employer or any of the Employer's Persons…"	√	√
7	"The carrying out by a Statutory Undertaker of work in pursuance of its statutory obligations in relation to the Works, or the failure to carry out such work"	√	
8	"Exceptionally adverse weather conditions"	√	
9	"Loss or damage occasioned by any of the Specified Perils"	√	
10	"Civil commotion or the use or threat of terrorism and/or the activities of the relevant authorities in dealing with such event or threat"	√	
11	"Strike, lock-out or local combination of workmen affecting any of the trades…"	√	
12	"The exercise after the Base Date by the UK Government of any statutory power … which directly affects the execution of the works"	√	
13	"Delay in receipt of any necessary permission or approval of any statutory body…"	√	√
14	"Force majeure"	√	

Contractors, at the commencement of or early in the course of a contract, is required to prepare and submit a Programme of the Works to the Contractor Administrator showing completion at a date materially before the contract date.

The Contract Administrator may approve such a programme or accept it without comment. It is then argued that the Contractor has a claim for damages for failure by the Contract Administrator to issue instructions at times necessary to comply with the programme.

6.7 CHANGE MANAGEMENT UNDER PMBOK

6.7.1 Procedure

Change management under the guidance of PMI (2013) includes five main stages, namely formulate change, plan change, implement change, manage transition, and sustain change, as shown in Figure 6.6.

During the formulate change stage, the project manager should identify the need for change, identify change resource, identify stakeholder expectation, coordinate change management with programme management, delineate scope of change, and start communication of the changes (PMI, 2013).

In the plan change stage, the activities to be carried out include the following:

- establish change approach and integrate change management activity with project programme and clarify associated risks
- plan stakeholder engagement and establish stakeholder communication model, and
- plan transition and integration

Then, in the third stage of implement change, the process starts with prepare change, then mobilize stakeholders, and finally deliver project outcome, including assess change process and modify the scope, activities, programme and budget.

During the manage transition stage outputs will be transitioned into the business, measure adoption rates and realized benefit.

Finally, the sustain change stage will undertake ongoing consultation with stakeholders, conduct sensemaking activities, and control benefit realization.

6.8 CHANGE THEME UNDER PRINCE2

Under PRINCE2, change management is undertaken through managing issues. The term "issue" in PRINCE2 (Axelos, 2017) refers to anything that would result in a change of the end product, in terms of time, cost, scope, quality, risk, and benefit.

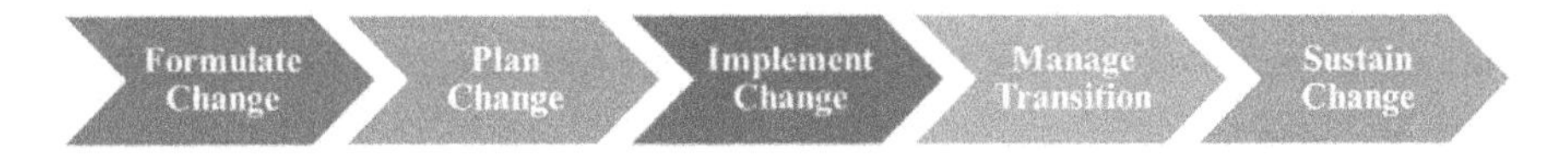

FIGURE 6.6 Project level change management process under PMI (2013).

The issues are classified into three types, namely request for change, off-specification, and a problem or concern.

The tools and techniques used in PRINCE2 change management include: change control approach, configuration item record, product status account, daily log, issue register, and issue report. The change management approach is developed during the "initiating a project" stage. It provides the procedure to undertake change control and issue management. As the project progresses, it will be reviewed and updated, e.g., during the "manage stage boundary" stage. The change authority and change budget are the key aspects for PRINCE2 change management. By default, the project board is the change authority, but it often delegates to the Project Manager to undertake day to day change decisions. Change budget is the allowance that the project sets aside to deal with potential changes; only the project board has the authority to allocate it.

The team manager raises any issues in concern of the Project Manager, and the Project Manager will manage the issues register and escalate when relevant tolerance is exceeded to the project board.

6.9 KEY ISSUES

The other key issues when managing changes in large construction projects are the condition precedent and time bar. Other key issues associated with the time at large and *quantum meruit* has already been discussed in Chapter 3 and Chapter 4 respectively.

6.9.1 Conditions Precedent

Under construction contracts, it is typical that the Contractor needs to submit a request to change to the Project Manager before undertaking any works different from the project scope provided by the contract. Although NEC allows the *Contractor* to undertake the works related to the compensation event before the implementation of the compensation event, the *Contractor* may expose themselves to the potential underestimated compensation event of the *Project Manager*'s assessment. Likewise, Sub-Clause 13.3.2 of FIDIC 2017 also requires that the Contractor does not delay any works while awaiting the Engineer's response to his proposal variations, and the Contractor is entitled to the payment of the cost it incurred related to submitting the proposal. However, the Contractor still bears the risks for not receiving full payment for the works it has undertaken.

In fact, the good practice set out by the contractor or consultant organization is actually not to commence works outside the contract until the change application has been approved by the Project Manager under NEC, the Engineer under FIDIC, or the Contract Administrator under JCT.

Although *Temloc Limited v Errill Properties Limited* (1987) 39 BLR 30 CA concludes that giving notice is not mandatory but a directory requirement, in practice, the Contractor is often barred from claiming time and money due to the failure to serve the required notice.

Under English common law, for a condition precedent to be enforceable, it needs to achieve two conditions, which were laid out in *Bremer Handelgesellschaft mbH v Vanden Avenue Izegem PVBA* [1978] 2 Lloyd's Rep 109. Firstly, the contract provision needs to express a precise time limit. Secondly, the contract needs to spell out the consequence for missing the timetable which will waive future rights. In *Multiplex Constructions (UK) Ltd v Honeywell Control Systems Ltd (No. 2)* [2007] EWHC 447 (TCC), the court held that notice of change is a condition precedent for the valid claim.

Both NEC and FIDIC set out the provisions to achieve the two conditions mentioned above. Therefore, under the FIDIC and NEC contracts, serving notification of change is the condition precedent of the following claim; whereas, the notification of change is not a strict condition precedent for claims under the JCT contract.

6.9.2 Time Bar

During the claim of extension of time, the parties should pay attention to the potential time bar for a valid claim. If the contract provision achieves the two conditions of requiring an enforceable condition precedent, failure to comply with a notice requirement may disentitle the possibility of claiming and triggering the "time bar" for the claim.

Clause 61.3 of NEC4 provides a time bar for the *Contractor*'s entitlement to claim the compensation event if the *Contractor* is obliged to provide the notification of the compensation event. Likewise, Sub-Clause 20.2.1 of the FIDIC 2017 contract requires notification within 28 days as the condition precedent for the Contractor to claim additional time or cost. In contrast, JCT does not apply a time bar to the Contractor's notification of delay, and Clause 2.24 requires the Contractor to notify Relevant Events within a reasonable time.

In *London Borough of Merton v Stanley Hugh Leach* (1985) 32 BLR 51, the court held that under the JCT 63 contract, the failure of the Contractor to serve the required notice did not bar its right to claim extension of time. Consequently, there are more delay claims and disputes toward the end of the project under JCT contracts.

In *Australian Development Corporation Pty Limited v White Constructions (ACT) Pty Ltd & Drs* (1996), the Contractor lost its right to claim extension of time as a result of failure to notify within a 30-day period as stipulated in the contract.

Likewise, in the Scottish case of *City Inn Ltd v Shepherd Construction Limited* [2003] Scot CS 146, due to the non-compliance of the notice requirements the Contractor waived the right to extension of time, and as a consequence of failure to complete by completion date, liquidation damages arose.

In the English case of *Steria Ltd v Sigma Wireless Communications Ltd* [2007] EWHC 3454 (TCC), Judge Stephen Davies stated:

> *one can see the commercial absurdity of an argument which would result in the Contractor being better off by deliberately failing to comply with a notice condition than by complying with it.*

Under FIDIC 2017 contract, the time bar applies to both the Employer's claim and the Contractor's claim. Sub-Clause 20.2.1 of FIDIC 2017 Red/Yellow Books requires the claiming party to serve a Notice of Claim within a 28-day time period. Failure to serve the Notice of Claim within four weeks results in the claiming party losing their entitlement to claim additional payment, extension of the Time for Completion, or Defects Notification Period (DNP).

Under NEC, the time bar only applies to the events for which the *Contractor* bears the responsibility to notify the *Project Manager*, as shown in Table 6.5. Clause 61.3 of NEC provides an eight-week limit for the events for which the *Contractor* is responsible to notify, unless the *Project Manager* should have notified such as a change in the Scope. If the *Contractor* does not submit the change request to the *Project Manager* for the additional works, they would not be entitled to additional cost and/or time.

6.10 RECOMMENDATIONS

In order to manage the changes in large construction projects, apart from complying with the contract's requirement to serve relevant notice and submit relevant documents, it is also important to establish the change management plan at the beginning of the project. Both PMI (2013) and PRINCE2 (Axelos, 2017) require preparation of the change management plan/approach as one of the documents that needs to be established at the project initiation stage.

In order to identify the potential changes in time, it is also important to establish the appropriate links between cost management and time management. If the resource is available, the author suggests the following:

1) Establish clear linkages between the project scope and work breakdown structure.
2) If possible, establish clear linkage between the work breakdown structure and project cost account as well as timesheet or expenses coding.
3) Make the cost account in the P6 cost and/or resource loading match the project financial cost account.
4) Establish linkage between the assumption register with the project scope, as well as the risk register.
5) Clearly understand the linkage between the risk register and issues register.
6) Relate all the early warnings, variations/compensation events in the P6 programme once they have been identified and implemented and keep separate activity for the accepted change event and pending change event.
7) For accepted change events, update the programme, the resource and/or cost loading in accordance with the implementation value or extension of time in the P6 programme immediately.
8) Undertake regular review and updates of the assumption register, the change register, issues register, risk register, programme progress, and actual cost occurred, in order for further analysis to be undertaken, which can then illustrate potential changes.
9) Maintain effective communication across whole project participants and stakeholders and discuss potential changes as soon as identified.
10) The Contractor should appoint specific personnel or a team to manage and control changes.

Change management is often closely connected with issue management. When managing and controlling changes under the NEC contract, CIMAR is often used to manage communication in relation to the early warning notice, compensation event, risk register, and relevant instructions under the *Project Manager* or the *Supervisor.* CIMAR is an effective web-based tool to facilitate all parties in the contract to comply with their duties to serve notice, submit documents, provide quotations, acceptant requests, etc. Likewise, for large construction projects under the FIDIC contract, it is also worth establishing a similar tool/system at the front-end of the project. In a simpler way, a comprehensive Excel spreadsheet with embedded macro can also deliver the fundamental benefit to facilitate the parties in the contract to keep on top of their duties within the time limit under the contract.

If the resource is available, it is also worth undertaking quantitative cost risk analysis and/or quantitative schedule risks analysis when making decisions on the major changes.

In summary, successful change management requires an effective change management procedure established at the front-end of the project, then supported with personnel with competent skills, effective tools, and systems, and ongoing management and control of the benefit realized after implementing the change.

REFERENCES

Book/Article

Axelos. 2017. *Managing Successful Projects with PRICE2.* 2017 Edition. Norfolk, UK: TSO (The Stationery Office) / Williams Lea Tag.

PMI. 2013. *Managing Change in Organizations – A Practice Guide.* Newtown Square, PA: Project Management Institute.

Contract

FIDIC. 1999. *Conditions of Contract for Construction.* 1st Edition. (1999 Red Book). Geneva, Switzerland: The Fédération Internationale des Ingénieurs-Conseils.

FIDIC. 2017. *Conditions of Contract for Construction.* 2nd Edition. (2017 Red Book). Geneva, Switzerland: The Fédération Internationale des Ingénieurs-Conseils.

FIDIC. 2017. *Conditions of Contract for EPC / Turnkey Project.* 2nd Edition. (2017 Yellow Book). Geneva, Switzerland: The Fédération Internationale des Ingénieurs-Conseils.

FIDIC. 2017. *Conditions of Contract for Plant & Design Build.* 2nd Edition. (2017 Silver Book). Geneva, Switzerland: The Fédération Internationale des Ingénieurs-Conseils.

NEC. 2013. *NEC3 Engineering and Construction Contract.* London, UK: Thomas Telford Ltd.

NEC. 2017. *NEC4 Engineering and Construction Contract.* London, UK: Thomas Telford Ltd.

Cases

Amec Building Company Limited v Cadmus Investment Company Limited (1996) 51 ConLR 105

Australian Development Corporation Pty Limited v White Constructions (ACT) Pty Ltd & Drs (1996)

Barnett v Chelsea & Kensington Hospital [1969] 1 QB 428

Bremer Handelgesellschaft mbH v Vanden Avenue Izegem PVBA [1978] 2 Lloyd's Rep 109
City Inn Ltd v Shepherd Construction Limited [2003] Scot CS 146
London Borough of Merton v Stanley Hugh Leach (1985) 32 BLR 51
Modern Engineering Ltd v Gilbert Ash Ltd [1974] A.C. 689
Multiplex Constructions (UK) Ltd v Honeywell Control Systems Ltd EWHC 447 (TCC)
Obrascon Huarte Lain SA v Her Majesty's Attorney General for Gibraltar [2014] EWHC 1028 (TCC)
Peak Construction Ltd v McKinney Foundations Ltd [1970] 1 BLR 111
Steria Ltd v Sigma Wireless Communications Ltd [2007] EWHC 3454 (TCC)
Temloc Limited v Errill Properties Limited (1987) 39 BLR 30 CA
Walter Lilly & Company Ltd v Mackay and another [2012] EWHC 1773 (TCC)

7 Alternative Dispute Resolutions

7.1 INTRODUCTION

Due to the complexity of large international construction projects, disputes often arise during the project implementation. Because of the broad range of participants in large construction projects, the local litigation is usually not favorable to the international participants, therefore alternative dispute resolutions are often used. This chapter introduces the well-known alternative dispute resolution mechanisms used in large international construction projects, e.g., mediation, adjudication, dispute board, and arbitration. It then explains the procedure and relevant approaches set out in the leading standard forms of contract, including FIDIC, NEC, and JCT. Finally, it concludes with recommendations for drafting dispute resolution clauses during the contract negotiation stage as well as the pitfalls and best practice during the post-contract implementation stage.

7.2 ALTERNATIVE DISPUTE RESOLUTION

Alternative dispute resolution (ADR) is used to resolve disputes outside the courts. It plays a significant role in construction projects. In particular, in large international construction projects, parties do not usually want to refer a dispute to a local court, unless it is expressly required by the Client. As described in Figure 7.1, there are many alternative dispute resolution methods that have been used in construction projects, including negotiation, mediation, conciliation, adjudication, dispute board, early natural evaluation, expert determination, and arbitration.

The enforcement of the ADR decision is a critical feature when parties choose the dispute resolution methods. The decision for negotiation, mediation, and conciliation are not all binding to the parties. Negotiation is undertaken between the senior manager or the representative from both parties. Negotiation is often the first ADR method that the parties undertake because it is cheaper and quicker. Most small disputes can be resolved between the parties through negotiation.

If the parties fail to resolve the dispute through negotiation, or the relationship is broken between the contract parties, mediation or conciliation is usually the second choice for the parties. Mediation or conciliation engages an independent third-party mediator or conciliator to facilitate the parties to resolve the dispute. However, like negotiation, the decision arrived from mediation or conciliation is not binding to the parties.

If the parties would like to choose an ADR approach which can deliver a binding decision, then they need to further consider whether the binding decision is interim or final. The ADR approaches providing an interim binding decision include adjudication, dispute board, conflict avoidance panel (CAP), and early neutral evaluation

Alternative Dispute Resolution

	Non-Binding		Binding
Parties Self Determination	• Negotiation • Mediation • Conciliation	Interim	• Adjudication • Dispute Adjudication Board • Conflict Avoidance Panel (CAP) • Early Neutral Evaluation (ENE)
3rd Party Determination	• Dispute Avoidance Board • Dispute Resolution Board	Final	• Arbitration • Med-Arb • Expert Determination

FIGURE 7.1 Means of alternative dispute resolution.

(ENE). If a party of the contract is dissatisfied with the decision arrived at the relevant ADR approach, it can then escalate the dispute to a further approach, which provides the final and binding decision, such as arbitration, Med-Arb, expert determination, and litigation.

This chapter focuses on a discussion of the alternative dispute resolutions provided under the leading standard forms contract, including mediation and conciliation, dispute board, adjudication, and arbitration.

7.3 MEDIATION

Mediation is a process where an independent and impartial third party is appointed as the mediator to assist the parties to reach a settlement of the matters in dispute. The parties need to sign a mediation agreement before carrying out the mediation. Mediation typically involves a mediator meeting the parties jointly to oversee the negotiation between the contract parties or meeting the parties separately to hear their grounds of the dispute and understand their position in a potential settlement. The mediator then facilitates the parties to understand each other's position of matters in dispute and subsequently achieve a settlement. The parties can potentially resolve almost any dispute by mediation, apart from dispute arising from tort. Such disputes will be determined by the court. Mediation can also render solutions, which would not be available under arbitration or litigation, e.g., additional remedial works (Furst, Ramsey, Hannaford et al., 2019).

As described in Figure 7.1, the agreement achieved through the mediation procedure is interim but not binding to the parties unless the parties both sign the settlement agreement. The mediator only assists the parties in settling their dispute through facilitated negotiation. Hence, the mediator cannot decide or impose any decision to the parties. The fundamental decision is made by the parties themselves.

The parties can start the mediation at any time. Mediation usually takes one day. For complex disputes, it may take up to one week. In addition, mediation can provide more formal legal decision over the parties' direct negotiation. Therefore, in commercial practice, parties often undertake mediation before referring the matters in dispute to adjudication, arbitration, or litigation.

Due to the advantage of the speed and cost effectiveness of the mediation process, mediation is increasingly considered by the parties in large construction projects as a dispute resolution strategy. It can speed up resolution of the dispute, and it can save the legal costs of a lengthy and more formal dispute resolution process.

Before 2018, there was no formal rules or procedures set out by the leading institute for dispute resolutions. However, as most mediators are trained by the Centre for Effective Dispute Resolution (CEDR), the tools and processes provided by the CEDR are well used in most commercial mediations in the UK. In 2018, the Chartered Institution of Arbitrators published the formal CIArb Mediation Rules (2018), which provides comprehensive rules to ensure an appropriate mediation process is undertaken.

Parties undertaking mediation should also adhere to the statutory requirements. In 2008, the EU Mediation Directive 2008/52/EC was enforced for civil and commercial mediations and all member states had to implement the directive by November 2010. Therefore, mediation undertaken within the European Union and the cross-border mediation engaged in by European Union member states need to comply with the EU Directive 2008/52/EC.

In the UK, before going to trial, the court will encourage the parties to attempt a mediation at the first Case Management Conference under CPR Part 1(4)(e). If a Party unreasonably refuses to mediate before the court trial, even if they won the case, they may be subject to cost sanctions in accordance with CPR 44.3(4). The losing party has the burden to prove the unreasonableness of the winning party's refusal.

7.4 ADJUDICATION

Adjudication is a popular alternative dispute resolution method used in construction projects, particular in jurisdictions where statutory adjudication is provided.

7.4.1 Statutory Adjudication

Statutory adjudication was first introduced under Section 108 of the Housing Grants Construction and Regeneration Act (HGCRA) 1996. In order to resolve the cash flow issues in the supply chain of the construction industry in the UK, statutory adjudication was introduced in a spirit of "pay first argue later." Section 108 of the HGCRA 1996 gives rights to the contract parties of construction projects to adjudicate at any time. In *Macob Civil Eng. Ltd v Morrison Construction Ltd* [1999] EWHC Technology 254, Dyson J describes statutory adjudication as follows:

> *The intention of Parliament in the Act was plain. It was to introduce a speedy mechanism for settling disputes and construction contracts on a provisional basis, and requiring the decision of adjudicators to be enforced pending the final determination of disputes by arbitration, litigation or agreement....*

Statutory adjudication only applies to construction contracts in England, Wales, and Scotland under Sections 104 and 105 of the HRCRA 1996. Section 105(1) clearly provides that the project qualifies the definition of the construction operations as follows:

(a) construction, alteration, repair, maintenance, extension, demolition or dismantling of buildings, or structures forming, or to form, part of the land (whether permanent or not);
(b) construction, alteration, repair, maintenance, extension, demolition or dismantling of any works forming, or to form, part of the land, including (without prejudice to the foregoing) walls, roadworks, power-lines, aircraft runways, docks and harbours, railways, inland waterways, pipe-lines, reservoirs, water-mains, wells, sewers, industrial plant and installations for purposes of land drainage, coast protection or defence;
(c) installation in any building or structure of fittings forming part of the land, including (without prejudice to the foregoing) systems of heating, lighting, air-conditioning, ventilation, power supply, drainage, sanitation, water supply or fire protection, or security or communications systems;
(d) external or internal cleaning of buildings and structures, so far as carried out in the course of their construction, alteration, repair, extension or restoration;
(e) operations which form an integral part of, or are preparatory to, or are for rendering complete, such operations as are previously described in this subsection, including site clearance, earth-moving, excavation, tunnelling and boring, laying of foundations, erection, maintenance or dismantling of scaffolding, site restoration, landscaping and the provision of roadways and other access works;
(f) painting or decorating the internal or external surfaces of any building or structure.

Meanwhile, Section 105(2) of the HGCRA 1996 also expressly excludes the works from the definition of construction contract as follows:

(a) drilling for, or extraction of, oil or natural gas;
(b) extraction (whether by underground or surface working) of minerals; tunnelling or boring, or construction of underground works, for this purpose;
(c) assembly, installation or demolition of plant or machinery, or erection or demolition of steelwork for the purposes of supporting or providing access to plant or machinery, on a site where the primary activity is—
(i) nuclear processing, power generation, or water or effluent treatment, or
(ii) the production, transmission, processing or bulk storage (other than warehousing) of chemicals, pharmaceuticals, oil, gas, steel or food and drink;
(d) manufacture or delivery to site of:
    (i) building or engineering components or equipment,
    (ii) materials, plant or machinery, or
    (iii) components for systems of heating, lighting, air-conditioning, ventilation, power supply, drainage, sanitation, water supply or fire protection, or for security or communications systems, except under a contract which also provides for their installation;
(e) the making, installation and repair of artistic works, being sculptures, murals and other works, which are wholly artistic in nature.

The statutory adjudication is a typical 28-day procedure although the parties can agree to extend it. The parties can refer to adjudication at any time. It is often used to resolve disputes in regard to interim payments, defects of work, delay or disruption, and associated extension of time.

The adjudicator's decision is interim binding and the winning Party can enforce the adjudicator's award at the Technology and Construction Court (TCC) in the UK. The parties have no right to appeal the adjudicator's decision. Even if the adjudicator makes any errors in their decision, the court will usually still enforce the adjudicator's award. However, if the adjudicator breaches their duty of natural justice, or there is no valid contract or dispute between the parties, the parties can appeal the adjudicator's decision in the Court of Appeal or Supreme Court in the UK.

Challenge of the adjudicator's decision often relies on natural justice and/or impartiality. Even though the adjudicator needs to make the decision in a short period, similar to the procedure challenge under Section 68 of the Arbitration Act 1996, the challenges of natural justice of the adjudicator's decision are rarely successful. For example, in *Carillion Construction Ltd v Devonport Royal Dockyard Ltd* [2005] EWCA Civ 1358, the Court of Appeal reaffirmed the adjudicator's decision.

Although it was initially intended to be used for quick resolution of short-term disputes to keep the project going, and the losing party can refer the case to arbitration or the court within four weeks of receiving the adjudicator's decision, over 95% of adjudication cases are settled without further escalation.

Following the success of adjudication in the UK, other countries are also introducing adjudication as a statutory right for the construction contract parties, for example, Australia, New Zealand, Singapore, Malaysia, South Africa, Ireland, Hong Kong, and the Isle of Man. In addition, some projects funded by the World Bank also provide adjudication as part of the dispute resolution provisions.

7.4.2 Non-Statutory Adjudication

The implementation of contractual adjudication can be traced back to 1988, where contractual adjudication is provided in the revision of the JCT Standard Form of Building Contract with Contractor's Design 1981.

Both NEC Option W1 and JCT forms of contract provide contractual adjudication as the second tier of their dispute resolution mechanism. In addition, a dispute adjudication board can also be considered as a form of non-statutory adjudication, which will be discussed further in Section 7.5.

Non-statutory adjudication is usually informal. The parties can decide their own rules and procedures for such adjudication. JCT requires that the contractual adjudication is undertaken in accordance with the Construction Industry Council (CIC) Model Adjudication Procedure. In addition, the parties can also use the following rules and procedures developed by various organizations:

- The Centre for Effective Dispute Resolution (CEDR) Rule (2015 edition)
- The Institute of Chemical Engineers Adjudication Rules "Grey Book" (fourth edition)

- The Technology and Construction Solicitors Association (TeCSA) Procedural Rules for Adjudication (version 3.2.1, 2015 edition)
- The Technology and Construction Bar Association (TECBAR) Adjudication Rules (2012 edition)

Like statutory adjudication, the decision concluded from the contractual adjudication process is interim and binding. The parties also have no rights to appeal, unless the adjudicator breaches his duty of impartiality or natural justice. Further detail on contractual adjudication will be discussed in the relevant standard forms in Section 7.7.

7.5 DISPUTE BOARD

Dispute board is a broad term that includes the dispute review board, dispute avoidance board, dispute resolution board, and dispute adjudication board. There is also the combined dispute board, e.g., the Dispute Avoidance and Adjudication Board (DAAB) under the FIDIC 2017 suite of contract.

Chern (2015) defines a dispute board as follows:

> *A dispute board or dispute review board (DRB) or dispute adjudication board (DAB) is a 'job-site' dispute adjudication process, typically comprising three independent and impartial persons selected by the contracting parties.*

The concept of dispute boards was initially developed in the construction industry in the United States in the 1960s, where a technical Joint Consulting Board was used to decide the parties' dispute in the Boundary Dam project in Washington. In 1975, the Eisenhower Tunnel project in Colorado in the United States was the first project to formally use the dispute board in the contract provision to replace the Engineering's adjudication. Shortly after, in 1981, the World Bank funded the El Cajón Dam project in Honduras which was the first to use the dispute review board in an international project.

Following the clear benefit illustrated by using the dispute board, in 1992, FIDIC provided the dispute review board in the supplement of the fourth edition of the 1987 Red Book. Three years later, in 1995, the International Bank for Reconstruction and Development (IBRD) of the World Bank required projects with funding over $50 million to use the dispute review board as one of the means of dispute resolution. In 1997, the Asian Development Bank (ADB) and the European Bank for Reconstruction & Development (EBRD) also made dispute boards a mandatory requirement for projects financed by them. In 1999, the first edition of the FIDIC rainbow suites introduced the standing dispute board in the Red Book and the ad hoc adjudication board in the Yellow and Silver Books. In 2004, the FIDIC Pink Book, MDB Harmonized edition, was published. The Pink Book is used for projects funded by the Multilateral Development Banks (MDBs) and the dispute adjudication board is mandatory for projects funded by MDBs. Also, in 2004, the International Chamber of Commerce published the first ICC dispute board rules, which was then updated in 2015. In 2006, nine other Multilateral Development Banks also introduced the dispute board in projects funded by them. In recent years, construction contracts have started to promote the practice of dispute avoidance. As a consequence, NEC4 introduced a new Option W3 for a standing dispute avoidance board in June 2017. Six months later, in December 2017, FIDIC also published its 2017 suites of contract, which provides the Dispute Avoidance and Adjudication Board for the Red, Yellow, and Silver Books.

A dispute board is a means of dispute resolution, which is usually formed at the commencement of the project to follow construction progress, encourage dispute avoidance, and assist the parties to resolve potential dispute in time during the project implementation.

Dispute boards typically comprise one or more independent and impartial professionals who are selected by the contracting parties. In World Bank projects, the number of panel members is determined based on the value of the project. If the estimated value of the project is more than US$50million, then three members of the dispute board are required. Otherwise, if the estimated project value is between US$10 million and US$50 million, the sole-member panel can form the dispute board. If the value of the project is less than US$10 million, then the dispute is more likely to be determined by an adjudicator. Table 7.1 explains the criteria used to select panel members of dispute boards under the rules of leading dispute resolution institutions, e.g., ICC and CIArb, and FIDIC contract.

Following the development of different forms of dispute board and recognition in large international construction projects, many institutions provide their specific rules for the dispute board. The well-applied rules for the dispute board include the ICC Dispute Board Rules (2015), the CIArb's Dispute Board Rules (2014), and the AAA's Dispute Resolution Board Guide. Furthermore, the dispute board is subsequently incorporated into the dispute resolution provisions of standard form contracts, e.g., FIDIC and NEC4.

The main reason for the popularity of the dispute board is its value for money. It generally costs less than 1% of the amount of total construction contract and generally it can resolve the parties' dispute in time. For example, it was used in the £14 billion Channel Tunnel project, which was the largest infrastructure project in Europe in the 1990s. In this project, 13 disputes were referred to the dispute board and 12 of them were resolved without further escalation. According to the Dispute Resolution Board Foundation (DRBF)'s Dispute Board Project Database, which records over 2,800 international projects that used the dispute board from 1975 to 2017, 81% of the disputes were resolved by the dispute board.

TABLE 7.1
Criteria for Selection of Dispute Board Members

ICC Rules	CIArb Rules	FIDIC
• nationality & residence • language skills • training, qualifications and experience • availability • ability to conduct the work	• fluent in contract language • experienced professional in the type of the works • contract interpretation	• fluent in contract language • experienced professional in the type of the works • contract interpretation

There are three types of dispute boards:

- Dispute avoidance board as provided under the NEC4 and FIDIC 2017 contracts
- Dispute review/resolution board which is the original means of dispute board and provides non-binding recommendations
- Dispute adjudication board which can provide binding decisions to the parties

7.5.1 Dispute Avoidance Board

The dispute avoidance board plays an important role in large construction project management. Both FIDIC and NEC set out the dispute avoidance board as a means of dispute resolution during the contract implementation. However, the dispute avoidance member is not obligated to provide their advice or recommendations, and such recommendations are not binding to the parties.

7.5.2 Dispute Review Board/Dispute Resolution Board (DRB)

The dispute review board originated from the United States. The American Arbitration Association (AAA) applies the dispute review board for dispute resolution. The ICC also recognizes the effectiveness of the dispute review board as a means for dispute resolution.

Similar to the dispute avoidance board, the decision of the DRB is not binding. It only works as a recommendation to both parties, and it is the parties' decision whether to comply with such recommendations.

7.5.3 Dispute Adjudication Board

The dispute resolution board also derives from the United States. In contrast to the dispute avoidance board and dispute review board, the decision of the dispute adjudication board is interim-binding. FIDIC 1999 suite provides the option of a standing dispute adjudication board in the Red Book or ad hoc dispute adjudication board in the Yellow and Silver Books.

The advantage of the dispute adjudication board or dispute resolution board include:

- prompt decisions when disputes arise
- maintain the working relationship between the contract parties
- good value for money
- provide a binding decision, and potentially final if no further objections
- achieve high settlements ratio

Whereas the disadvantages of the dispute adjudication board or dispute resolution board are as follows:

- standing dispute resolution/adjudication board can be expensive
- decision is not final

- potential jurisdictional issues associated with the appointment or changing of DAB panel members

Although the standing dispute board will incur substantive cost, it is an effective dispute resolution mechanism compared to the general cost in large international projects.

7.6 ARBITRATION

Arbitration is usually the final stage for the parties to resolve disputes under an alternative dispute resolution route due to the complex nature of large international construction projects. The Contractor, the Subcontractor, and the Suppliers engaged in large international construction projects often come from different countries. For example, the Chinese Xiaolangdi Multi-purpose Dam project engaged Consultants, Contractors, Subcontractors, and Suppliers from 52 countries back in the 1990s. With increasing globalization over the past decades, many large international construction projects have parties participating from a number of different countries. The parties of the main Contractor Joint Venture may also come from different counties, for example the consortium of Line 1 of the Riyadh Metro project in Saudi Arabia come from the United States (Bechtel), Saudi Arabia (Almabani), Germany (Siemens), and Greece (CCC). The main Contractors for other lines of the project also include Bombardier from Canada, and Alstom from France. For such major state infrastructure projects, the international Contractors are concerned that the local courts may be biased towards its own country, therefore, the Contractors prefer to resolve disputes by arbitration instead.

7.6.1 Arbitration vs. Litigation

Because both arbitration and litigation can provide the final and binding decision, when choosing the last procedure to determine the dispute, the parties need to consider the advantages and disadvantages of each approach and the specific project features during contract drafting stage. There are many advantages of arbitration over the local court.

First, arbitration is confidential; not only the judgement but the case details are also confidential in arbitration. Information relating to the parties' dispute is not published to the public, whereas court cases are often published on the internet and everyone can read the details about the background of the project as well as the parties' disputes.

Second, a great feature of arbitration is enforcement within the 161 contracting states of the New York Convention (NYC). Therefore, the winning party can seek compensation from the losing party's assets in any of the 161 contracting states under the New York Convention. This provides the international contractor with more certainty when things go wrong, in order to obtain damages that they are entitled to.

Third, arbitration is undertaken by the arbitrator appointed either by the parties or a leading arbitration institution. The arbitrator is less likely to be under pressure from the local government or society and is consequently able to make more impartial decisions compared to the local court. In addition, international arbitration is usually undertaken according to the standard rules established by the leading arbitration institutions, for example, the ICC (International Chamber of Commerce), SCC (Stockholm Chamber of Commerce), LCIA (London Court of International Arbitration), CIArb (Chartered Institute of Arbitrators), AAA (American Arbitration Association), and ICSID (International Centre for Settlement of Investment Disputes). For ad hoc arbitration, the United Nations Commission on International Trade Law (UNCITRAL) model law is often used. In addition, some leading engineering institutions have also set out their arbitration rules or procedures, for example the ICE's (Institute of Civil Engineers) Arbitration Procedure, the Institution of Chemical Engineers' Pink Book, Arbitration Rules, and the Society of Construction Arbitrator's Construction Industry Model Arbitration Rules (CIMAR). The contractor, the subcontractor, or the suppliers engaged in large international construction projects do not need to understand the local laws for each project they undertake. By using the standard rules and procedures of arbitration provided by leading arbitration institutions, they feel more confident to handle the process and obtain a fairer and more enforceable award.

Fourth, as described in Figure 7.1, the arbitration award is final and binding. If a party fails to comply with the arbitration award, they can seek enforcement of the arbitration award in the court of one of 161 countries of the New York Convention.

Fifth, the competence of arbitrators is usually of a high quality in large international construction projects compared to other alternative dispute resolution approaches. The arbitral tribunal often constitutes arbitrators from different jurisdictions and expertise. Some retired judges and technical experts with competent legal qualifications also act as arbitrators in the arbitral tribunal. Therefore, the arbitral tribunal can usually apply the relevant applicable law and legal principles to make a fair decision.

7.6.2 Applicable Law

In order to obtain a binding and enforceable arbitration award, it is important to understand the applicable laws during the arbitration procedure. The source of law for international arbitrations is wide-ranging. The first source is the international conventions. The most important convention for international commercial arbitration is the 1958 New York Convention. For investor-state arbitration, the 1965 Washington Convention together with the relevant bilateral investment treaties (BIT) are generally used. In addition, the 1971 European Convention and the 1975 Panama Convention are also well-known conventions applied in international arbitration. Second, international "soft" laws are often used in international arbitration, for example the UNCITRAL Model Law, the IBA's (International Bar Association) rules and guidance in relation to conflict of interest and taking evidence. Third, the arbitration rules have been developed by arbitration institutions or engineering institutions as discussed in Section 7.6.1. Forth, the transnational principles and *lex mercatoria*, e.g., the UNIDROIT Principles of International Commercial Contracts may also be used. In addition, the national case law in the common law jurisdictions can also constitute the source of law for the arbitration.

The governing law of international arbitration is an important aspect that the parties should consider before signing the contract. Unlike domestic projects, international arbitration may involve multiple applicable laws in a single case. In *Channel Tunnel Group v Balfour Beatty Construction* [1993] 1 Lloyd's Rep 291, [1993] AC 334, Lord Mustill stated that multi-national laws might apply to an international arbitration. Therefore, the law regulating the main contract could be different from the law governing the arbitration agreement, and/or the law of the seat. The five laws typically governing international arbitration include the following:

- the *lex fori*: the substantive law governing the main contract
- the *lex arbitri*: the law governing the arbitration procedure
- the law governing the arbitration agreement
- the law governing the enforcement of the arbitral award, and
- the law applicable to the parties' capacity

The substantive law is the law of contract, which subsequently governs the dispute. For example, all standard forms of contract provides the Contractor the rights to claim for change of law. If English law is selected to govern the main contract, then English case law as well as the relevant statute will be applied, e.g., the implied terms of satisfactory quality, fitness for purpose, or reasonable skill and care under the Sales of Goods Act (SGA) 1979 and the Supply of Goods and Services Act (SGSA) 1982.

The law of the seat is the law governing the arbitration procedure. London and Paris are often chosen as the seat of arbitration. Consequently, English law or French law will govern the arbitration procedure, which means the English court or the French court can undertake emergency arbitration, hear challenges to the jurisdiction or award, and/or grant anti-suit injunctions.

The law governing the party's capacity is also important. If the contract party is found to have no capacity under the applicable law in the country where the project is undertaken, then the arbitration agreement will be void as well as the arbitration award.

Under the New York Convention, the winning party can seek enforcement in any of the 161 countries registered with the New York Convention. However, it is important for the party to understand the law of the country in which they are intending to seek efficient enforcement. For example, in *Yukos Universal Limited (Isle of Man) v. The Russian Federation*, UNCITRAL, PCA Case No. AA 227, Yukos won the arbitration and was awarded compensation of US$50 billion, which is the largest arbitration award in the world. However, the arbitration award was set aside by the Hague District Court in the Netherlands as the consequence of the jurisdiction challenge by the Russians. Nevertheless, Yukos continuously commences enforcement proceedings in the UK, Germany, France, and the United States.

The most complicated, but most important, component is the law of arbitration agreement, because without a valid arbitration agreement, the arbitral tribunal has no jurisdiction to decide the dispute. Even if the arbitral tribunal rendered the arbitration award, the risk associated with the validity of the arbitration agreement can void the arbitration award.

In practice, the arbitration clause in construction contracts does not provide a governing law clause. The FIDIC contract simply provides that arbitration shall be conducted under the ICC rules. Article 21.1 of the ICC Arbitration Rules (2017) provides that:

> *The parties shall be free to agree upon the rules of law to be applied by the arbitral tribunal to the merits of the dispute. In the absence of any such agreement, the arbitral tribunal shall apply the rules of law which it determines to be appropriate.*

The Contract Particulars of the FIDIC 2017 suite of contract does not provide detailed provisions for arbitration, e.g., seat of arbitration, or any governing law. In the absence of the express provisions, the ICC rules will apply to the appointment of the arbitrator; the seat of arbitration will be Paris; and the law governing the arbitration agreement will be decided under the ICC rules (Tan and Mian, 2017). Likewise, JCT 2016 forms of contract require the use of the JCT 2016 Construction Industry Model Arbitration Rules (CIMAR). Consequently, the substantive law, the *lex arbitri*, the enforcement law, as well as the law governing the arbitration agreement are all the law of England and Wales.

If the parties use a bespoke contract, in the absence of clear indication of the law governing the arbitration agreement, uncertainty may arise. Traditionally, the English court applies the substantive law approach. In *Sonatrach Petroleum Corp v Ferrell International* [2002] 1 All E.R. (Comm) 627, Mr Justice Colman stated that:

> *Where the substantive contract contains an express choice of law, but the agreement to arbitrate contains no separate express choice of law, the latter agreement will normally be governed by the body of law expressly chosen to govern the substantive contract.*

Following the application of the Arbitration Act 1996, the English court shifted the emphasis of the law governing the arbitration agreement towards the law of the seat. Later on, in *SulAmérica Cia Nacional de Seguros S.A. v. Enesa Engenharia S.A.* [2012] 1 Lloyd's Rep 671, CA, the English court set out a three-stage test to determine the law governing the arbitration agreement, which is express choice, implied choice, and the closest connection. If the arbitration clauses do not expressly provide the law regulating the arbitration agreement, the law governing the substantive contract will be the implied choice as a presumption of the arbitration agreement. This presumption may be rebutted with further facts. If there is no governing law provision in the main contract, the law of the seat will govern the arbitration agreement.

The *SulAmérica* approach has also been adopted in other jurisdictions. For example, in *Klöckner Pentaplast Gmbh v Advance Technology* HCA1526/2010, the Hong Kong court held that in the circumstances that there is no expressed provision for the governing law of the substantive contract, the court will apply the law of the main contract to the arbitration agreement. In contrast, the Singapore court takes a different view from the English court's decision in *SulAmérica* and leant towards implementing law of the seat as the law governing the agreement to arbitrate. In *FirstLink Investments Corp v GT Payment Ptel* [2014] SGHCR 12, the judge considered that when the parties' relationship breaks down during the dispute, they would prefer a neutral approach and choose the law governing the arbitration procedures for the arbitration agreement rather than the law of the substantive contract. In addition, the importance of arbitral seat is illustrated in both Articles 34(2)(a)(i) and 36(1)(a)(i) of the UNCITRAL Model Law and Article V(1)(a) of the New York Convention.

Therefore, the parties should take careful consideration of the applicable law of the arbitration in accordance with the relevant jurisdictions. It would be helpful to clearly specify the applicable laws in the arbitration clause.

7.7 DISPUTE RESOLUTION IN STANDARD FORMS

Standard forms of construction contract generally provides multi-tier dispute resolution provisions in order to resolve the dispute between the parties in a timely and efficient manner.

Figure 7.2 demonstrates the typical three tiers of dispute resolution mechanism in the contract of large construction projects. When a dispute arises, the senior management from both parties commence negotiation of the matters in dispute. If the parties fail to resolve the dispute, then the amicable settlement, which includes mediation, adjudication, dispute resolution board (DRB), dispute adjudication/avoidance board (DAB), or conflict avoidance panel (CAP) will usually be undertaken. If the parties still cannot resolve the dispute, then the dispute will be referred to arbitration or litigation. In the alternative dispute resolution approach under tier two and tier three, the parties may challenge the decision, seeking enforcement from the court, when the litigation procedures may be engaged. In case of the parties' failure to challenge the decision made in tier two adjudication or dispute board within the time limit set out in the contract, the decision will then be final and binding. Both NEC and FIDIC made amendments to the dispute resolution provisions in their latest edition published in 2017.

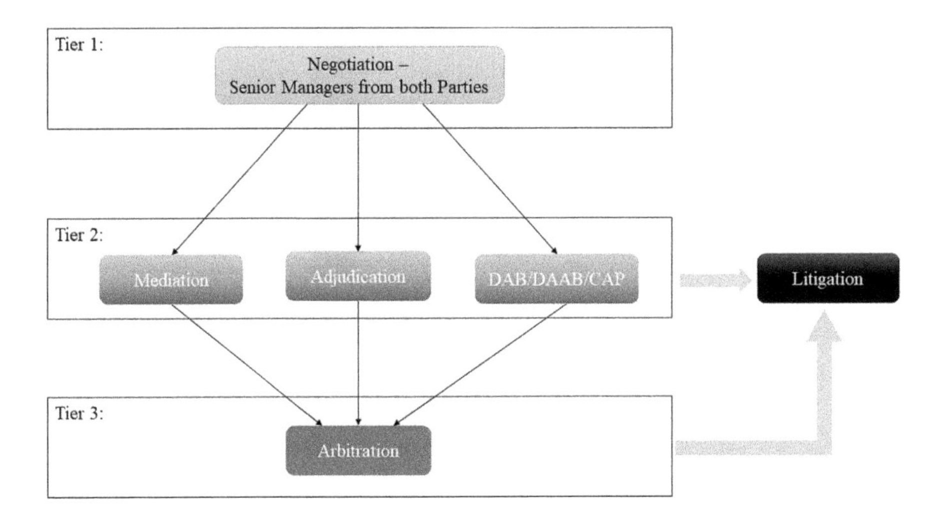

FIGURE 7.2 Multi-tier dispute resolution in large construction projects.

7.7.1 FIDIC

FIDIC 2017 edition separates the dispute resolution provisions from Clause 20 in the FIDIC 1999 edition and creates a new Clause 21 "Dispute ad Arbitration." Clause 21 prescribes a multi-tiered dispute resolution procedure:

1) submission of a claim to the Engineer, who makes a determination in accordance with Sub-Clause 3.5
2) referral of the dispute to the Dispute Adjudication Board (DAB)
3) the giving of a notice of dissatisfaction with the DAB's decision
4) amicable settlement, and
5) arbitration

Figure 7.3 illustrates the dispute resolution mechanism under the FIDIC 2017 contract. There are three main tiers of the dispute resolution procedures under FIDIC 2017 contract, namely the Engineer's decision, the Dispute Avoidance/Adjudication Board (DAAB)'s decision, and arbitration.

7.7.1.1 Tier 1 – Engineer's Determination

When a dispute arises, the Engineer will first consult the Parties to encourage the Parties to discuss the matters in dispute and negotiate a settlement agreement in accordance with Sub-Clause 3.7.1 within 42 days. If the Parties reach an agreement

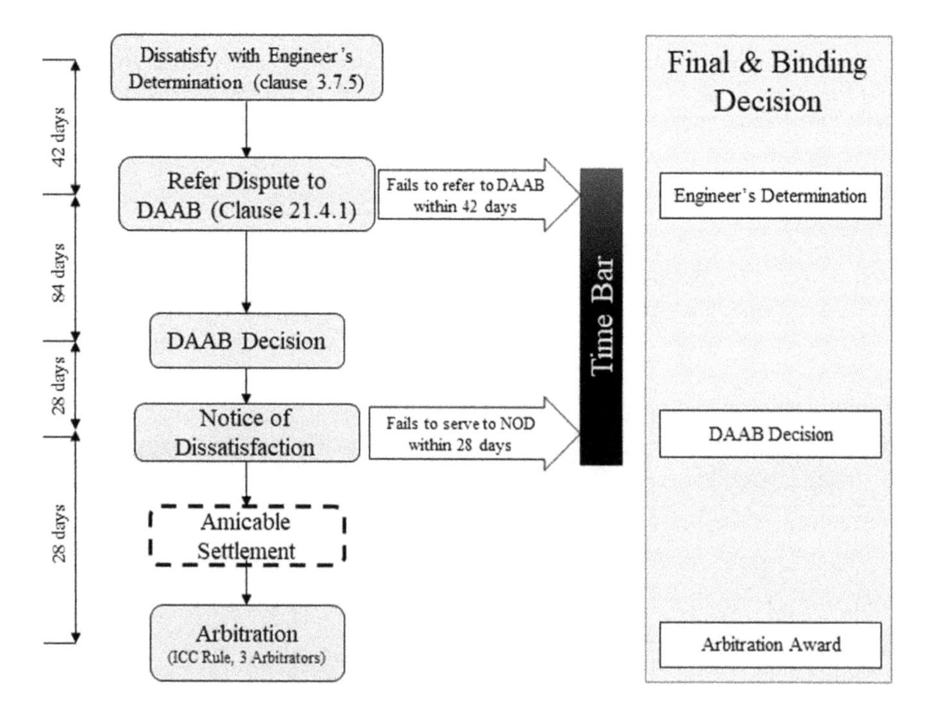

FIGURE 7.3 Dispute resolution mechanism under FIDIC 2017 contract.

within 42 days, then the Engineer will give the Notice of agreement in accordance with Sub-Clause 3.7.3.

However, if the Parties cannot achieve an agreement within 42 days, or both Parties inform the Engineer that no agreement can be achieved within 42 days, then the Engineer will give Notice to the Parties when whichever these two situation arise earlier and proceed the Engineer's Determination by giving a "Notice of the Engineer's Determination" in accordance with Sub-Clause 3.7.2 within 42 days.

If either Party does not satisfy the Engineer's Determination, the Engineer should issue a notice of dissatisfaction (NOD) within 28 days under Sub-Clause 3.7.5. The Parties should then refer the NOD to the DAAB within 42 days. If the Parties fail to refer the NOD within 42 days, then the time bar applies and the Engineer's Determination on the issues in dispute will be final.

7.7.1.2 Tier 2 – Dispute Avoidance and Adjudication Board

FIDIC 1999 suite of contracts provides two types of dispute adjudication board, which is the standing dispute adjudication board in the FIDIC 1999 Red Book and the ad hoc dispute adjudication board in the FIDIC 1999 Yellow and Silver Books. Like other leading construction contracts, FIDIC 2017 also emphasis its intention to improve dispute avoidance. Therefore, the new concept of a dispute avoidance board is introduced in addition to the dispute adjudication board in the FIDIC 1999 suite.

The Parties need to appoint the panel member/s of the DAAB by the date stated in the Contract Data or within 28 days of receiving the Letter of Acceptance. The sole-panel member needs to be selected from the name list in the Contract Data. If a three-member panel is applied, each Party needs to appoint one member and then both Parties agree the chair member after consulting with the two appointed members. The panel member/s need to sign a DAAB Agreement, from then the DAAB is formed. The member of the DAAB may be replaced with the consent of both Parties. The Parties will share the cost of the DAAB.

When potential dispute arises, the Parties can jointly request the DAAB to provide suggestion of the resolution in accordance with Sub-Clause 21.3, although the DAAB does not have the obligation to provide such a suggestion. The DAAB's informal advice at this stage is not binding.

Either Party can refer the dispute to the DAAB for a formal decision in accordance with Sub-Clause 21.4.1. Following the reference of a dispute to the DAAB, both Parties need to provide all information to the DAAB unless the contract has already been suspended or terminated. The DAAB then needs to issue the written decision to the Parties and the Engineer within 84 days or a period agreed by both Parties. If either Party does not satisfy the DAAB's decision, the Engineer needs to serve a notice of dissatisfaction within 28 days. In contrast to the DAAB's suggestion under Sub-Clause 21.3, the DAAB's decision under Sub-Clause 21.4 is binding to both Parties. If both Parties fail to serve a notice of dissatisfaction within 28 days, the DAAB's decision will be final and binding.

Failure to comply with the DAAB's decision will constitute a breach, and the other Party can further refer to arbitration under Sub-Clause 21.7.

Dispute resolution via the dispute avoidance and adjudication board is the condition precedent for the Parties to refer the dispute to arbitration. The parties can only bypass the DAAB if no DAAB is in place under Sub-Clause 21.8.

The FIDIC 2017 edition also provides the General Conditions of Dispute Avoidance/Adjudication Agreement as an Appendix.

7.7.1.3 Tier 3 – Arbitration

Unlike the NEC and JCT, because FIDIC is drafted aiming for international construction projects, arbitration is the primary choice for the Parties to conclude the dispute. Therefore, litigation does not appear as an option in the last stage of its multi-tier dispute resolution clauses.

Sub-Clause 21.6 provides that the Parties can only refer the dispute to arbitration in the following circumstances:

1) Dissatisfaction of both the Engineer's Determination under Sub-Clause 3.7.5 and the DAAB's decision under Sub-Clause 21.4.4, or
2) Dissatisfaction of both the Engineer's Determination under Sub-Clause 3.7.5 and No DAAB in place under Sub-Clause 21.8

Following the notice of dissatisfaction of the DAAB's decision, it is optional for the Parties to undertake the amicable settlement negotiation within 28 days of NOD under Sub-Clause 21.5. If the dispute cannot be resolved within 28 days, then either Party can refer the matters in dispute to arbitration. FIDIC requires arbitration to be undertaken by three arbitrators under the ICC Rules of Arbitration.

Unless agreed by both parties otherwise, arbitration under the FIDIC contract should apply the Rules of Arbitration of the International Chamber of Commerce (ICC). The appointment of arbitrator/s should comply with the ICC Rules of Arbitration. Sub-Clause 1.4 sets out the language for arbitration.

The Parties should also decide the seat of arbitration, which governs the procedure of arbitration. For example, if the seat of arbitration is London, then English law will apply. Consequently, the Arbitration Act 1999 will apply, and the English court has the power to issue an "anti-suit injunction" and grant emergency relief.

The ICC Arbitration Rule is the default arbitration rule used in the FIDIC contract. It expressly waives the parties' rights to appeal under Article 35.6, which provides that (ICC, 2017):

> *Every award shall be binding on the parties. By submitting the dispute to arbitration under the Rules, the Parties undertake to carry out any award without delay and shall be deemed to have waived their right to any form of recourse insofar as such waiver can validly be made.*

Therefore, even if the seat of arbitration is defined as London and the English law applies, the parties cannot appeal error of point of law under Section 69 of the Arbitration Act 1996, because it is an optional provision. However, the Parties can still appeal under both Section 67 for error of jurisdiction and Section 68 for error

TABLE 7.2
NEC4 Dispute Resolution Options

Tier	Option W1	Option W2	Option W3
1	Negotiation – Senior Representative	Negotiation – Senior Representative	
2	Contractual Adjudication	Statutory Adjudication	Dispute Avoidance Board (DAB)
3	Litigation or Arbitration	Litigation or Arbitration	Litigation or Arbitration

of procedure under the Arbitration Act 1996, as both Section 67 and Section 68 are mandatory provisions.

Furthermore, Sub-Clause 21.5 provides that the arbitrator/s have the power to open up, review and revise any certification, determination, instruction, valuation of the Engineer, and the decision of the DAAB.

7.7.2 NEC

NEC3 provides two dispute resolution options, which are the contractual adjudication under Option W1 and the statutory adjudication under Option W2 for the first tier and litigation or arbitration for the second tier. NEC4 introduces an additional initial negotiation stage before adjudication under Option W1 and Option W2. NEC4 ECC forms of contract also introduces a new Option W3 to provide a solution through a standing dispute avoidance board (DAB), aiming to shift the focus of dispute resolution in NEC3 to dispute avoidance in NEC4. The structure of the dispute resolution options under the NEC4 contract is explained in Table 7.2.

Across all options, undertaking the second tier process is the condition precedent to proceed the tier three dispute resolution procedure. Therefore, the Parties cannot refer the dispute to litigation or arbitration until the adjudication or DAB fails to resolve the dispute. Under Option W1, the first-tier negotiation is also the condition precedent for the contractual adjudication. However, under Option W2, because the Parties have the statutory rights to proceed adjudication at any time, the negotiation is not a condition precedent to statutory adjudication.

7.7.2.1 Tier 1 – Negotiation

In order to incorporate the wide range of good practice in dispute resolutions and encourage dispute avoidance, which is reflected the change of title to "*Resolving and Avoiding Disputes*" in NEC4 (2017) from "*Dispute Resolution*" in NEC3 (2013), NEC4 adds a new stage of negotiation between the senior representatives from both Parties. Clause W1.1 requests the referring Party to notify a dispute in accordance with the "Dispute Reference Table" to the *Senior Representatives* within four weeks for the following matters under Clause W1.1.(4):

- Either Party can refer for action/inaction of the *Project Manager* or the *Supervisor*.

- The *Client* can refer the dispute for a programme, compensation event, or quotation for the compensation event after it has been accepted or deemed accepted under Clause 66.2.
- Either Party can refer a dispute of assessment of Defined Cost after it has been treated as correct when the *Project Manager* issues the Payment Certificate.

Furthermore, either Party can refer a dispute to the adjudicator or tribunal when disputes arise.

For the first three circumstances where the claiming Party refers the dispute to the *Adjudicato*r or the *tribunal*, the notification to the *Senior Representatives* is a condition precedent for the referring Party. Failure to notify within four weeks results in its entitlement to dispute under Option W1 being time barred. Because the dispute resolution needs to follow on from the procedure set out in multi-tier dispute resolution provisions, if the dispute has been time barred in the first-tier negotiation stage, the Party would not be allowed to pursue the latter on adjudication in the second tier, and arbitration or litigation in the third tier. Consequently, failure to comply with the four weeks' notice period under Clause W1.1 will waive the claiming Party's rights to dispute the relevant issues.

In contrast, Clause W2.1 does not provide a time limit for the referring Party to notify the *Senior Representatives.*

Figure 7.4 explains the process of the tier one "resolving disputes" under Clause W1.1 and W2.1 of the NEC4 contract.

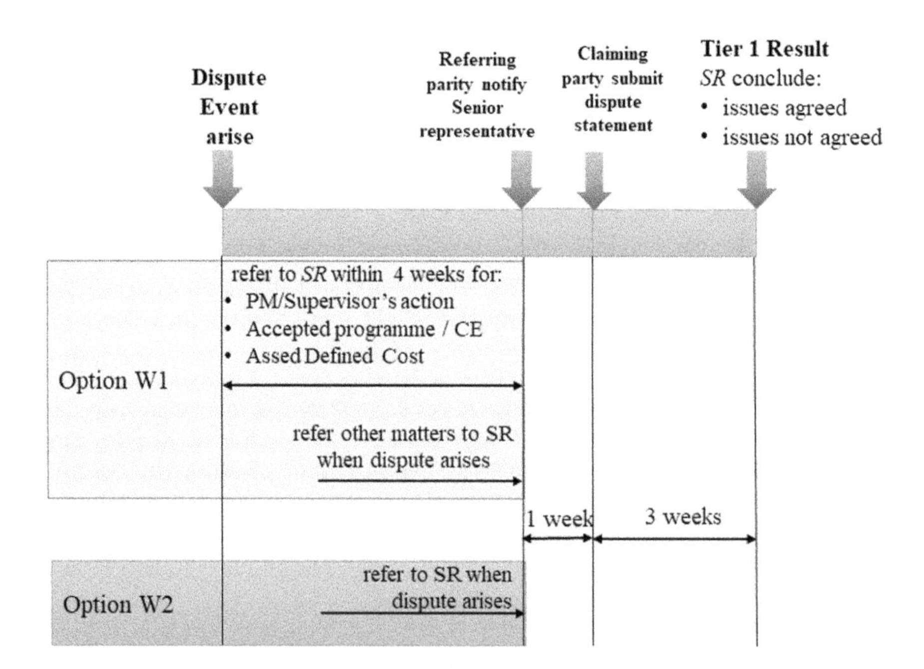

FIGURE 7.4 NEC4 tier one "resolving disputes" under Option W1 and Option W2.

Both Options W1 and W2 require the claiming Party to serve the statement of case and supporting evidence limited to ten A4 pages within one week of referral, then the *Senior Representatives* attend necessary meetings to discuss the dispute and conclude with a list of agreed issues and un-agreed issues within three weeks. If the Parties fail to achieve agreement with the *Senior Representatives*, the referring Party should serve notice of adjudication within two weeks, and the dispute needs to be referred to the adjudicator within one week of the notice of adjudication. The *Adjudicator* then has four weeks to make their decision on the matters in dispute and inform both Parties and the *Project Manager.*

7.7.2.2 Tier 2 – Adjudication/Dispute Avoidance Board

The second tier of the dispute resolution structure constitutes three options, which are contractual adjudication under Option W1, statutory adjudication under Option W2, and dispute avoidance board under Option W3.

7.7.2.2.1 Option W1 – Contractual Adjudication

The Parties need to appoint an *Adjudicator* at the starting of the contract. In the absence of indicating the *Adjudicator*'s name in the Contract Data, both Parties should appoint an *Adjudicator* jointly. Otherwise, the *Adjudicator nominating body* will appoint an *Adjudicator* within seven days of referral.

Clause W2.2 requires the *Adjudicator* to act impartially. Failure to comply with the impartiality under Clause W1.2(2) may result in the challenge of natural justice of the *Adjudicator*'s award. As a consequence, the *Adjudicator*'s decision will be void.

Figure 7.5 explains the contractual adjudication process under Option W1. Either Party can refer the issues that haven't been agreed by the *Senior Representatives* to adjudication by issuing a notice of adjudication to the other Party and the *Project Manager* within two weeks after concluding the tier one results under Clause W1.3(1). The two-week time limit can be extended by the agreement of both parties, but if the referring Party fails to refer the dispute within the time limit required in the contract, both Parties are time barred from referring the list of unagreed issues to the *Adjudicator* or the tribunal. The referring Party then needs to refer the dispute to the *Adjudicator* within one week after the notice of adjudication and submit relevant information to the *Adjudicator* and the other Party within four weeks of referral.

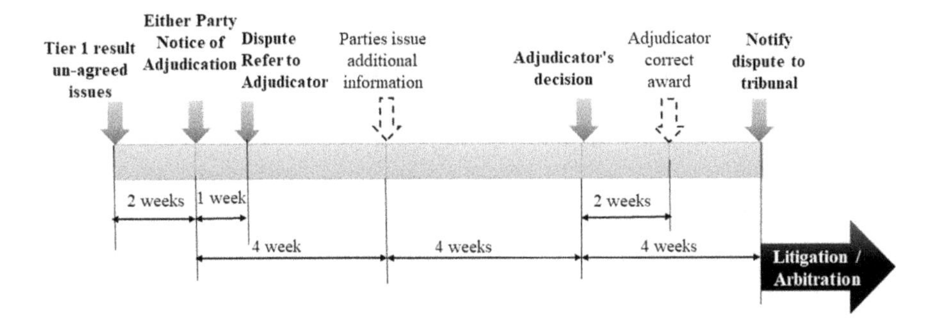

FIGURE 7.5 Contractual adjudication process under NEC4 Option W1.

The *Adjudicator* must give his decision to both Parties and the *Project Manager* within four weeks of receiving the last information from the referring Party. The *Adjudicator* can correct his decision within two weeks of informing of the award. The *Adjudicator*'s decision is binding but interim. If either Party is not satisfied with the *Adjudicator*'s decision, he can then notify the other Party of the intention to refer the specified issues to the tribunal within four weeks of the *Adjudicator*'s decision. If the Parties fail to refer the dispute to the tribunal within four weeks, the *Adjudicator*'s decision will be final and binding and both Parties are time barred from referring the dispute to the tribunal.

7.7.2.2.2 Option W2 – Statutory Adjudication

Option W2 applies to construction projects within the UK, or the countries that provide similar statutory adjudication. The definition of construction projects is provided in Sections 104 and 105(1) of the Housing Grants, Construction and Regeneration Act (HGCRA) 1996 as discussed in Section 7.4.1. Section 108 of the HGCRA 1996 then gives the Parties the statutory rights to dispute at any time.

In accordance with Clause W2.2(3), the Parties need to appoint the *Adjudicator* at the commencement of the project under the NEC Dispute Resolution Service Contract. Similar to Option W1, in the absence of an identified *Adjudicator* in the Contract Data, the Parties can appoint an *Adjudicator* jointly or selected by the *Adjudicator nominating body.*

Figure 7.6 explains the procedure for the statutory adjuration under Option W2 of the NEC4 contract.

The referring Party needs to send a notice of adjudication to the other Party and the *Adjudicator* identified in the Contract Data. The *Adjudicator* needs to decide whether they can decide the dispute within three days of receiving the notice of adjudication. In the case of failure to decide within three days, the *Adjudicator* is deemed to be resigned.

Otherwise, the referring Party needs to refer the dispute to the *Adjudicator* within seven days and provide relevant information to the *Adjudicator* and the other Party. The *Ajudicator* needs to inform both Parties of his decision and the *Project Manager* within 28 days of referring. The decision period may be extended for a further 14 days with the consent of the referring Party, or other period agreed by both Parties

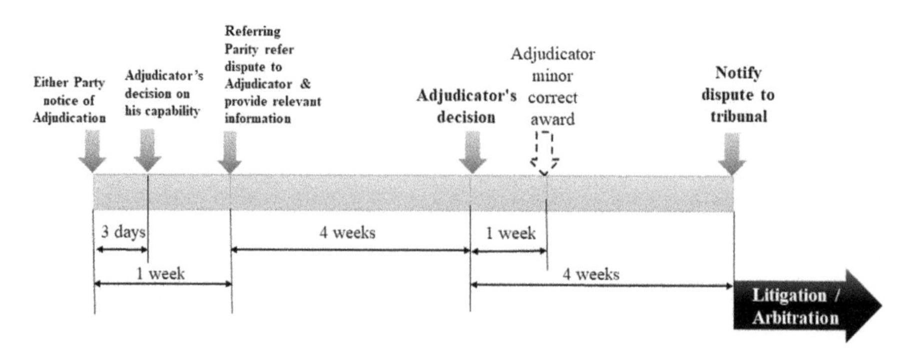

FIGURE 7.6 Statutory adjudication procedure under Option W2 of NEC4 contract.

under Clause W2.3(8). If the *Adjudicator* fails to inform his decision within 28 days or the extended period, the *Adjudicator* is treated as resigned. Like Option W1, the *Adjudicator* can make minor corrections within five days of his award.

The *Adjudicator*'s decision is interim binding to the Parties unless it has been further referred to the tribunal. Similar to Option W1, if either Party is dissatisfied with the *Adjudicator*'s decision, he need to notify the other Party of his intention to refer the dispute to the tribunal. Failure to notify within four weeks waives both Parties' rights to refer to the tribunal and the *Adjudicator's* decision becomes final and binding.

7.7.2.2.3 Option W3 – Dispute Avoidance Board

NEC4 introduces the *Dispute Avoidance Board* (DAB) as a new alternative dispute resolution mechanism under Option W3 to facilitate the Parties to resolve "*potential dispute before they become dispute*" (NEC, 2017).

The DAB constitutes one or three panel members. The one-member panel is usually a specialist in the specific type of project undertaken and with other competent knowledge, e.g., legal, commercial, or planning. The three-member panel usually includes a lawyer, a technical expert, and a quantity surveyor or a project manager.

Similar to Option W2, the Parties need to appoint the DAB panel member/s under the NEC Dispute Resolution Service Contract at the start of the project. In the absence of any member specified in the Contract Data, the Parties can appoint any panel member jointly. Otherwise, the *Dispute Avoidance Board nominating body* will appoint the required panel member within seven days in accordance with Clause W3.1(4).

The dispute board panel is required to meet or visit the Site in intervals set out in the Contract Data. Figure 7.7 explains the dispute avoidance procedure under Option 3 of the NEC4 contract.

When a potential dispute arises, either Party can notify the other Party and the *Project Manager*. Either Party then needs to refer the potential dispute to the *Dispute Avoidance Board* within two to four weeks. The Parties then need to submit the contract, progress report, and other relevant information to the DAB. The DAB will visit the Site, undertake investigations, review relevant documents, and then provides the DAB's recommendations to resolve the potential dispute.

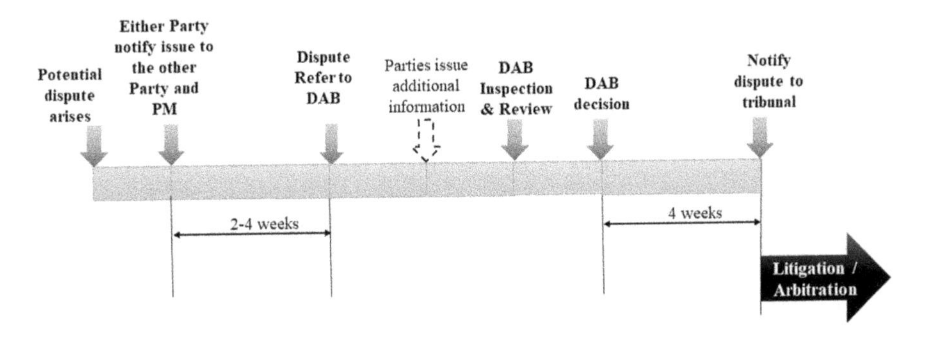

FIGURE 7.7 Dispute avoidance board procedure under Option W3 of NEC4 contract.

Unlike Option W1 and Option W2, there is no strict time limit set out for the process of the DAB. Because the purpose of the DAB is to assist resolving the potential dispute, the DAB's recommendation is not binding to the Parties. The DAB process is more intended to provide the Parties as well as the DAB panel with more flexibility to resolve the potential disputes more effectively.

Although the decision of the DAB is not binding, like Options W1 and W2, it is mandatory for the Parties to refer to the DAB before referring to the tribunal. Under Clause W3.3(2) of NEC4, if the Party does not give dissatisfactory notification of the DAB's recommendations within four weeks, both Parties will be time barred from further referring to the tribunal. Consequently, the DAB's decision will become final but not binding. This creates uncertainty for the Parties in dispute. In particular, if one Party does not comply with the DAB's recommendations when arbitration and litigation has been time barred, whether the other Party can then refer the matters in dispute to the tribunal is uncertain. Therefore, for large, complex international construction projects, Parties should take additional consideration of provisions under Option W3 and make necessary amendments during the contract negotiation stage.

7.7.2.3 Tier 3 – Tribunal

The tribunal includes arbitration or litigation under the NEC4 contract. The tribunal can overturn the *Adjudicator*'s decision. As discussed at the beginning of Section 7.7.2, the contractual adjudication under Clause W1.4(1), the statutory adjudication under Clause W2.4(1), and the *Dispute Avoidance Board* under Clause W3.3(1) are the condition precedent for the Parties' to further dispute with litigation or arbitration. Litigation or arbitration cannot proceed without adjudication or the DAB under the NEC4 contract.

Due to the success of statutory adjudication in the UK, 95% of disputes decided by the adjudication are concluded. Therefore, since the application of the Housing Ground and Construction Regeneration Act 1996, domestic arbitration cases in the UK have reduced significantly.

Furthermore, the Technology and Construction Court (TCC) deals with cases efficiently and at relatively low cost. Therefore, domestic construction disputes in the UK primarily use adjudication as the preliminary formal legal procedure and litigation for the final decision.

If the Parties intend to refer the dispute to the local court, then the civil procedure in the relevant jurisdiction should be followed.

If the Parties intend to reach the final decision through arbitration, then they should specify arbitration rules to be used, the seat of arbitration, and the procedure to appoint the arbitrator/s in the Contract Data.

7.7.3 JCT

JCT provides a multi-tier dispute resolution mechanism which is set out in Articles 7, 8, and 9 of the Contract Particulars and Section 9 of Conditions of the contract. The process starts with voluntary direct negotiation as the first tier. Then in the second tier, the Parties can choose voluntary mediation; otherwise statutory or contractual adjudication will apply. If the Parties fail to resolve the dispute from adjudication, then the dispute will be escalated to arbitration or litigation in the third tier.

7.7.3.1 Tier 1 – Negotiation

JCT does not set out clear provisions for negotiation, but Clause 9.1 refers to direct negotiation before the mediation. Item 6 "Notification and negotiation of disputes" of Schedule 8 "Supplemental Provisions", the eighth recital in JCT SBC 2016, provides that (JCT, 2016):

> *With a view to avoidance or early resolution of disputes or differences, each Party shall promptly notify the other of any matter that appears likely to give rise to a dispute or difference. The senior executives nominated in the Contract Particulars (or if either is not available, a colleague of similar standing) shall meet as soon as practicable for direct, good faith negotiations to resolve the matter.*

The negotiation is not a mandatory requirement to proceed alternative means of dispute resolution, but it is recommended that the Parties undertake negotiation before initiating any formal dispute resolution procedures.

7.7.3.2 Tier 2 – Mediation/Adjudication

Voluntary mediation and adjudication constitute the second tier of the JCT dispute resolution mechanism.

7.7.3.2.1 Mediation

Like negotiation, mediation is neither a mandatory procedure required under the JCT contract, nor a condition precedent to proceed further dispute resolution procedures. However, the JCT contract recommends that the Parties take thoughtful consideration to let a third-party mediator to facilitate the Parties to resolve the dispute. Clause 9.1 provides that (JCT, 2016):

> *Subject to Article 7, if a dispute or difference arises under this Contract which cannot be resolved by direct negotiations, each Party shall give serious consideration to any request by the other to refer the matter to mediation.*

Clause 43.1 of the JCT Major Project Construction Contract 2016 (MPCC) provides more specific provision for mediation as follows (JCT, 2016):

> *Either Party may identify to the other any dispute or difference as being a matter that he considers to be capable of resolution by mediation and, upon being requested to do so, the other Party shall within 7 days indicate whether or not he consents to participate in a mediation with a view to resolving the dispute. The objective of mediation under clause 43 shall be to reach a binding agreement in resolution of the dispute.*

As mediation is increasingly used in the dispute resolution and has achieved substantial success, if the Party refuses to undertake mediation without a persuasive reason, it may be subject to cost sanctions under Article 44.4(3) of the Civil Procedure Rules (Minister of Justice, 2019), where the English court encourages the Parties to mediate the dispute as advised in paragraph 165 of the JCT DB 2016 Guide.

7.7.3.2.2 Adjudication

As JCT contracts are mainly used in the UK, Section 108 of the Housing Grants, Construction and Regeneration Act (HGCRA) 1996 automatically give the Parties the statutory rights to resolve the dispute with adjudication. If the construction contract does not comply with the requirements under Sections 104 and 105 of the HGCRA 1996, then the Scheme for Construction Contracts (England and Wales) Regulations 1998 ("the Scheme") will supplement the provisions of the contract which are lacking or replace the non-compliance provisions to enable the construction contract to comply with the HGCRA 1996.

Article 7 and Clause 9.2 provide relevant provisions for adjudication under the JCT contract. JCT refers the adjudication in accordance with the Scheme, which provides provisions for adjudication under Part I of the Scheme. Therefore, the adjudication under the JCT contract is mainly governed by the Scheme. If the project is outside the definition of the construction project under Section 105 of the HGCRA 1996, the statutory adjudication would not apply. Therefore, JCT provides the contractual adjudication with similar rights to the statutory adjudication for projects that cannot proceed with statutory adjudication under the Scheme.

The Parties need to agree the Adjudicator and specify his name in Article 9.2.1 of the Contract Particulars. In the absence of a specific Adjudicator's name, the Parties should choose a nominating body from the following list:

- Royal Institute of British Architects
- Royal Institution of Chartered Surveyors
- constructionadjudicators.com
- Association of Independent Construction Adjudicators
- Chartered Institute of Arbitrators

Clause 9.2.2 requires disputes to be opened up for inspection or tested under Clause 3.13.3 of the JCT DB 2016 and 3.18.4 of JCT SBC 2016 contract; if the adjudicator does not have relevant technical competence, they need to appoint an independent expert to advise and provide a written report.

Similar to NEC, JCT also provides its own Adjudication Agreement 2016 for the Parties to appoint the Adjudicator. In addition, the JCT Adjudication Rules govern the adjudication procedure for homeowner/occupier contract forms. For instance, it requires the Adjudicator to give his written decision within 21 days after his appointment; it also sets out the maximum hourly rate of £150 for the adjudicator and caps the charge time to 15 hours.

7.7.3.3 Tier 3 – Arbitration/Litigation

Like NEC, the final tier of the dispute resolution under the JCT contract is arbitration or litigation, which provides the final and binding decision.

7.7.3.3.1 Litigation

Due to the effective performance of the Technology and Construction Court (TCC) in the UK, and JCT is mainly used in the UK, litigation is the default final means to resolve the dispute under JCT forms of contract. JCT does not provide specific provisions for litigation, but it will follow the Civil Procedure Rules (CPR) and relevant court guidance in the UK.

7.7.3.3.2 Arbitration

If the Parties decide to use arbitration for the final resolution, then the institution to nominate the Arbitrator needs to be specified in Article 9.4.1 of the Contract Particulars. The Parties should choose a President or a Vice-President of following list of institutions to appoint the Arbitrator:

- Royal Institute of British Architects
- Royal Institution of Chartered Surveyors
- Chartered Institute of Arbitrators

In accordance with Article 8, all disputes can be concluded with arbitration, apart from issues relating to the Construction Industry Scheme, VAT, or enforcing the adjudicator's decision will be decided in the court.

If the Parties chose arbitration as the final means to resolve the dispute, the JCT Construction Industry Model Arbitration Rules (CIMAR) 2016 will apply under Clause 9.3. Clause 9.4 then provides that if either Party intends to refer the dispute to arbitration, he need to serve a notice of arbitration to the other Party in accordance with Rule 2.1. The Parties then need to agree the Arbitrator within 14 days, otherwise the President or a Vice-President of the Parties' selected institution in Article 8 will appoint the Arbitrator in accordance with Rule 2.3. In the conflict of multiple arbitral proceedings under different arbitration agreements, Clause 9.4.2 refers to Rule 2.6, 2.7, and 2.8.

Clause 9.5 provides the power of the Arbitrator as follows:

> *to rectify this Contract so that it accurately reflects the true agreement made by the Parties, to direct such measurements and/or valuations as may in his opinion be desirable in order to determine the rights of the Parties and to ascertain and award any sum which ought to have been the subject of or included in any payment and to open up, review and revise any account, opinion, decision, requirement or notice and to determine all matters in dispute which shall be submitted to him in the same manner as if no such account, opinion, decision, requirement or notice had been given.*

Clause 9.6 emphasizes that the Arbitrator's award is final and binding, unless it is successfully challenged under point of law in Clause 9.7.

Unlike the FIDIC contract, which excludes the Parties' right of appeal on error of law, Clause 9.7 expressly provides the Parties' right to appeal question of law under Section 69 of the Arbitration Act 1996.

7.8 RECOMMENDATIONS

Disputes raised in large construction projects can be long-lasting, expensive, and complex. Furthermore, too frequently disputes in a project can have a detriment impact on the relationships between the contracting parties. When the relationship is broken, collaboration, one of the key success factors of large construction projects, is also eliminated. Consequently, it has a further negative impact on the project. This section provides recommendations for the Parties to consider during both the pre-contract and post-contract stages, in order to avoid unnecessary disputes and resolve the inevitable disputes more efficiently and effectively.

7.8.1 Pre-Contract

The notion of dispute avoidance has attracted more and more attention in recent years. Both the JCT 2016 suite and NEC4 contract emphasize the concept of dispute avoidance and provide additional provisions in their new edition published in 2017. In order to fundamentally avoid disputes, the contract Party should take effective action from the contract drafting and negotiation stage, rather than dealing with disputes when they arise.

In the pre-contract stage, the contract Parties should ensure that the contract is finalized between both Parties and signed before commencing the works. Quite often, the construction contract starts with a letter of intent. However, once the works have started, and the interim payment is provided, the Parties loss the momentum to get the final contract signed. However, if things go wrong and disputes arise, the Contractor will request payment on the basis of *quantum meruit*, whereas the Employer would argue that there is no formal contract in place. Therefore, to avoid such disputes in the future, it is better to have the formal contract signed by both contract Parties before starting the work.

One of the reasons leading to most disputes in construction contracts is the Employer's change of the scope. As discussed in Chapter 5, due to the complexity and length of large construction projects, changes are inevitable. The inconsistency or ambiguity of the scope document can also cause confusion between the contract parties, which then leads to delay in approving changes, and consequently delaying the project. Therefore, it is important that the scope of work is clearly defined in the project. Consequently, the changes can be identified and evaluated appropriately.

The Contractor's works not achieving the satisfied quality is another reason that often gives rise to disputes in large construction projects. Likewise, it is important to set out clear provisions for the required quality for each component of works in the contract as the benchmark to evaluate the acceptance of the works when completed.

It is also important to set reasonable provisions in dealing with changes, as change events often arise as a consequence of the uncertainty embedded in the project, which is difficult to estimate. It is a good practice for the NEC contract to evaluate all compensation events based on the defined cost. As a consequence, the works undertaken outside the scope will be paid on a reimbursable basis, which will eliminate the dispute among the Parties in regard to the valuation of change events.

Before entering into a construction contract, the Parties should also ensure that the contract provisions are fair and clear, particularly for bespoke contracts. It is recommended to use the standard forms of contract, which has already been implemented and considered by a broad range of legal and technical experts over decades. When undertaking any revisions of the standard forms of contract, it is strongly suggested that the Parties seek advice from the competent legal personnel who are familiar with the relevant standard forms of contract and inter-relationship between the contract clauses. Inappropriate revision of the standard forms of contract may cause inconsistency or ambiguity within the contract, which may give rise to disputes during the implementation.

It is also important to balance the risk allocation between the contract Parties during the contract drafting stage. NEC allocates the risks to the Parties that can best handle it. It is not only a fair approach, but also an effective approach when considering the overall benefit to the whole project. Unbalanced risk allocation would lead to potential problems during contract implementation. Consequently, this gives rise to disputes.

Early warning is a good feature which contributes to the success of the NEC contract. As a consequence, FIDIC 2017 also provides a new provision for early warning. It is good practice to include provisions for early notification of potential issues and set out appropriate procedures to incentivize contract Parties to follow the process.

It is also worth setting up effective contract management tools prior to the project commencement. For example, installing the communication tool CIMAR for projects under the NEC contract and having the relevant staff trained in advance.

Due to the complexity and long duration of large construction projects, it is not possible to completely avoid disputes. It is important for the contract Parties to consider dispute resolution strategies during contract negotiation and the drafting stage. It is recommended to have multi-tier dispute resolution provisions in the contract, which provide the contract Parties with the ability to resolve small disputes in a more effective manner in terms of both time and cost. Negotiation should be the first tier of dispute resolution. In accordance with the specific nature of the construction project, mediation, adjudication, or various forms of dispute board can be used for the second tier of dispute resolution. Finally, arbitration or litigation, which provide the final and binding decision, should be the last tier of dispute resolution.

In addition, when using the FIDIC contract in jurisdictions providing statutory adjudication (e.g., offshore wind projects in the UK), the Parties should consider the potential conflict between the statutory adjudication and the DAAB provided in the FIDIC contract and make necessary amendments before signing the contract.

As the arbitration clause in construction contracts is generally silent to the seat of the arbitration, as well as the law governing the arbitral procedures, this can create additional conflict between the Parties in dispute. The choice of law can have a significant impact on the international arbitration. In particular, if a bespoke contract is used, it is important for the Parties to specify the seat of the arbitration and the law governing the arbitration agreement. The Parties should carefully consider the applicable law throughout the arbitration procedures and make clear provisions in the contract because in the absence of express provisions to the applicable law, additional issues may arise in the international arbitration.

7.8.2 Post-Contract

Once the project commences, it is important to undertake appropriate contract management and control. Because large construction projects usually take many years, the key personnel may change during the project implementation, and it is important to keep clear records of early notifications, change records, communication records, and inspection records.

Under both the FIDIC and NEC contracts, time bar and condition precedent have been used in a number of clauses. It is important for the contract Parties to comply with the action required and time limit in the contract.

When disputes arise post contract incorporation, the Parties should work collaboratively to avoid formal dispute resolution procedures. Even if it is not expressly required by the contract, when disputes arise, it is important for the senior management or representative from both Parties to review the dispute and negotiate a potential resolution before using any formal dispute resolution procedures. If the contract provides mediation (e.g., JCT forms of contract), then the Parties should take the opportunity and work together with the mediator to settle the dispute. Likewise, the Parties should consider the recommendations provided by the dispute avoidance board if available and try to settle the matters in dispute at an early stage.

REFERENCES

Book/Article

Chern, Cyril. (2015). *Chern on Dispute Boards: Practice and Procedure*. 3rd edition. London, UK: Informa Law from Routledge.

CIArb. 2014. *Dispute Board Rules*. London, UK: The Chartered Institute of Arbitrators.

CIArb. 2018. *CIArb Mediation Rules*. London, UK: The Chartered Institute of Arbitrators.

DRBF. DB Project Database. http://www.drb.org/publications-data/drb-database/

Furst, S., Ramsey, V., Hannaford, S. et al. 2019. Chapter 17 Arbitration and ADR. *Keating on Construction Contracts*. 10th Edition. London, UK: Sweet & Maxwell.

ICC. 2015. *Dispute Board Rules*. Paris, France: International Chamber of Commerce.

ICC. 2017. *Arbitration Rules*. Paris, France: International Chamber of Commerce.

JCT. 2016. *Design and Build Contract Guide (DB/G)*. London, UK: The Joint Contracts Tribunal Limited.

Ministry of Justice of the UK. 2019. The *Civil Procedure Rules*. https://www.justice.gov.uk/courts/procedure-rules/civil, Accessed on 18 November 2019.

Tan, Y. and Mian, S. 2017. China-Pakistan Economic Corridor (CPEC) Dispute Resolution Mechanism. *TDM (Transnational Dispute Management) Journal*, 3. Netherlands: Maris B.V.

Contract

FIDIC. 1999. *Conditions of Contract for Construction*. 1st Edition. (1999 Red Book). Geneva, Switzerland: The Fédération Internationale des Ingénieurs-Conseils.

FIDIC. 2017. *Conditions of Contract for Construction*. 2nd Edition. (2017 Red Book). Geneva, Switzerland: The Fédération Internationale des Ingénieurs-Conseils.

FIDIC. 2017. *Conditions of Contract for EPC / Turnkey Project*. 2nd Edition. (2017 Yellow Book). Geneva, Switzerland: The Fédération Internationale des Ingénieurs-Conseils.

FIDIC. 2017. *Conditions of Contract for Plant & Design Build*. 2nd Edition. (2017 Silver Book). Geneva, Switzerland: The Fédération Internationale des Ingénieurs-Conseils.

JCT. 2016. *Design and Build Contract 2016*. London, UK: The Joint Contracts Tribunal Limited.

JCT. 2016. *Major Project Construction Contract 2016*. London, UK: The Joint Contracts Tribunal Limited.

JCT. 2016. *Standard Building Contract with Quantities 2016*. London, UK: The Joint Contracts Tribunal Limited.

NEC. 2013. *NEC3 Engineering and Construction Contract*. London, UK: Thomas Telford Ltd.

NEC. 2017. *NEC4 Engineering and Construction Contract*. London, UK: Thomas Telford Ltd.

STATUTE

EU Mediation Directive (2008/52/EC).

Housing Grants, Construction and Regeneration Act 1996 (HGCRA 1996).

Scheme for Construction Contracts (England and Wales) Regulations 1998 (the Scheme 1998).

CASE

Carillion Construction Ltd v Devonport Royal Dockyard Ltd [2005] EWCA Civ 1358.

Channel Tunnel Group v Balfour Beatty Construction [1993] 1 Lloyd's Rep 291, [1993] AC 334.

Klöckner Pentaplast Gmbh v Advance Technology HCA1526/2010.

Macob Civil Eng. Ltd v Morrison Construction Ltd [1999] EWHC Technology 254.

Sonatrach Petroleum Corp v Ferrell International [2002] 1 All E.R. (Comm) 627.

SulAmérica Cia Nacional de Seguros S.A. v. Enesa Engenharia S.A. [2012] 1 Lloyd's Rep 671, CA.

Yukos Universal Limited (Isle of Man) v. The Russian Federation, UNCITRAL, PCA Case No. AA 227.

8 Integrated Project Control

8.1 INTRODUCTION

A successful project is evaluated based on whether the project can deliver the satisfactory end product on time and within the budget. Therefore, time, cost, and quality are recognized as the "golden triangle" of project management. In order to achieve project success, project management professionals are increasingly realizing the importance of integrated project management. In particular, establishing appropriate integrated project control at the front-end of the project can make a big difference to the end result of a project. This chapter begins by introducing a brief sample of an integrated project control system from the pre-contract stage to the post-contract implementation. It then introduces the components of the project control in detail, including earned value management, schedule control, cost control, change control, risk control, and progress reporting. Next, the chapter explains how to retain continuous improvement after establishing bespoke integrated project control for a project, particularly by using key performance indicators and lean techniques. Finally, recommendations for undertaking effective project control are provided.

8.2 INTEGRATED PROJECT CONTROL SYSTEM

In accordance with the 2019 global project control survey report undertaken by Logikal (2019), the fully integrated project control system can improve the overall project success rate by up to 89%, which is five times more than if the project was undertaken without the integrated project control system. Therefore, the executive fully integrated project control system and process can significantly improve the project success rate.

8.2.1 Integrated Project Control

This section briefly introduces how to use an example of an effective integrated project control system to undertake effective management and control from the tendering stage, to the beginning of the project, and throughout the project implementation.

8.2.1.1 Pre-Contract

The integrated project control starts as early as the tendering stage. Many projects fail because of inaccurate cost estimation during the tender stage. If the organization has appropriate tools and system in place, it would be better to calculate the total project cost based on the bottom-up estimation.

However, in an environment with a high degree of competition, sometime the senior management of the organization briefly estimate a tender price that can win the bid and implement the top-down estimation approach. In practice, the bottom-up estimation usually ends up about 20% higher than the top-down estimation. The top-down estimation focuses more on the absolute components required for the project but may ignore some key risks associated with the project.

When the tender information is received, the project manager should organize an interactive workshop of all technical leaders to review the scope of work, agree the work breakdown structure (WBS) and delegate to the accountable owner for each WBS element.

For each WBS component, a standard Excel template can then be used for capturing the activity name, required resource name, and number of units required for each resource identified for each activity. Meanwhile the assumptions made for the estimation and associated risks are also captured in the same Excel template.

All the spreadsheet input from the WBS owners is then consolidated by the project control team. At this stage a project database or comprehensive Excel spreadsheet functioning as a simple database that can be established, which includes the following fields:

- WBS ID
- WBS Name
- Activity ID
- Activity Name
- Resource Name
- No. of Units
- Duration
- Expenses

Then based on the background rate associated with each resource, the additional following fields can be calculated:

- Bare Rate
- Sell Rate
- Revenue
- Cost
- Profit Margin

In addition, a separate sheet can be generated for assumption and risks respectively including the following fields:

- WBS ID
- WBS Name
- Assumption/Risk ID
- Assumption/Risk Description
- Activity ID (if available)

As the standard estimating template is used, an Excel VBA tool can be developed to automatically consolidate all information and produce a database for cost estimating, together with a list of assumptions and a list of risks. This can be very useful for later stage as regards project control, as the activity ID, the activity name, and the duration can be imported into the planning software, e.g., Primavera P6, to automatically generate the activities with required duration within the programme. Once the logic has been established within the activities, the schedule is complete. A separate spreadsheet containing the activity ID, resource name, resource rates, and number of units can then be imported into the programme to complete the resource loading. Because P6 can calculate the cost automatically by using the resource rate multiplied by the number of units, once the resource loading is complete, with a few further amendments, the cost loading can also be completed. This can significantly improve the programme efficiency under a tight tender duration.

8.2.1.2 Post-Contract: Front-End Tool and System

If the bid is successful, immediately after the project is awarded, an interactive planning workshop should be arranged. This workshop ideally engages with the project director, project manager, discipline leaders from both the Contractor and the Client. Depending on how large and complex the project is, such an interactive planning workshop may take half a day to a couple of days. A short-term and long-term wall chart can be used to facilitate the team to develop the integrated programme. It usually starts with a long-term chart to agree the high-level programme and then shifts to the short-term chart to develop the detailed programme based on the key dates agreed in the high-level programme.

The planner will then prepare the programme in accordance with the output of the planning workshop and circulate it to all participants for comments. Then the planner incorporates comments in the revised programme and issues for the Client's approval. Because the programme is developed jointly with both contracting parties and all potential interface parties, the final submitted programme is likely to be accepted quickly by the Client. This provides the opportunity to have a contract baseline programme shortly after the project commences. The agreed baseline programme provides the foundation for later stage as regards change management, cost control, and schedule control. Therefore, it plays a crucial role in project control as well as project success. In particular for projects undertaken in the traditional method, where the baseline programme can only be updated with the agreed change events, it is even more important to have it in place as early as possible.

During the interactive planning workshop, the assumptions and risks identified will also be recorded. Then the assumptions register and the risk register will be updated accordingly to facilitate change management and risk management later on.

The resource and cost loading can be justified within the new programme. During this stage, it is also important to consider how to set up the cost control account. In order to conduct the earned value management more effectively, the cost control account can be made more consistent with the project programme, e.g., the work breakdown structure, the discipline, work packages, etc. There are three ways to establish the linkage between the activities in the programme and cost records in the financial system as follows:

1) If the cost account is consistent with the WBS, then there is no additional works required.
2) If the cost account is not consistent with the WBS but is consistent with other programme elements, e.g., discipline, location, or work packages, and each activity only associates to a single cost account, the linkage can be established for the relevant activity code in the programme and the programme layout can be set up to switch the programme into the view as per cost account.
3) In the worst-case scenario, if a single activity is associated with multiple cost accounts, then the cost account needs to be established in Primavera P6 and for each resource allocation, the cost account needs to be allocated accordingly.

The distribution of the cost allocated to each resource can also be defined during cost allocation in order to reflect the realistic resource profile for each activity.

Once the resource and cost loading are completed in the programme, then the resource and cost profile for the project life cycle is available to set up on the project reporting template. The progress report usually includes the executive summary for the project progress in the period and summarizes the key activities in the following period, the status for the key milestones and key activities, the status of cost to date and the forecast outturns, the status of change events, the top risks and potential mitigations, the key quality management indexes, and the S-curve for the earned value management.

Furthermore, as the planner prepares the baseline programme, the quantity surveyors should finalize the resource and cost allocation, the risk manager should prepare the risk register, the change manager should prepare the assumptions register and issues register, and the project manager or the project control manager should prepare the project control procedure and ideally published that within one month of project commencement.

Due to the time bar and condition precedent clauses in the standard forms of contract, it is also important to develop a system or install the relevant software to facilitate contract management and communication management at the front-end of the project.

8.2.1.3 Post-Contract: During Implementation

Following establishing of the integrated project control system, all project participants are required to comply with the project control procedure during the implementation of the project. Therefore, the contract management team will facilitate the project manager, construction manager, and section managers to identify changes in the project and submit relevant notice as required by the contract. The change manager will keep records for the communication associated with each change event and the status to date and then report in the monthly progress report.

In each period, the planners update the progress of the programme with the actual percent complete as well as the actual Start/Finish date and forecast Start/Finish date for each activity. The quantity surveyors capture the actual cost for the project to date and project forecast cost. The risk manager holds the risk review meeting and

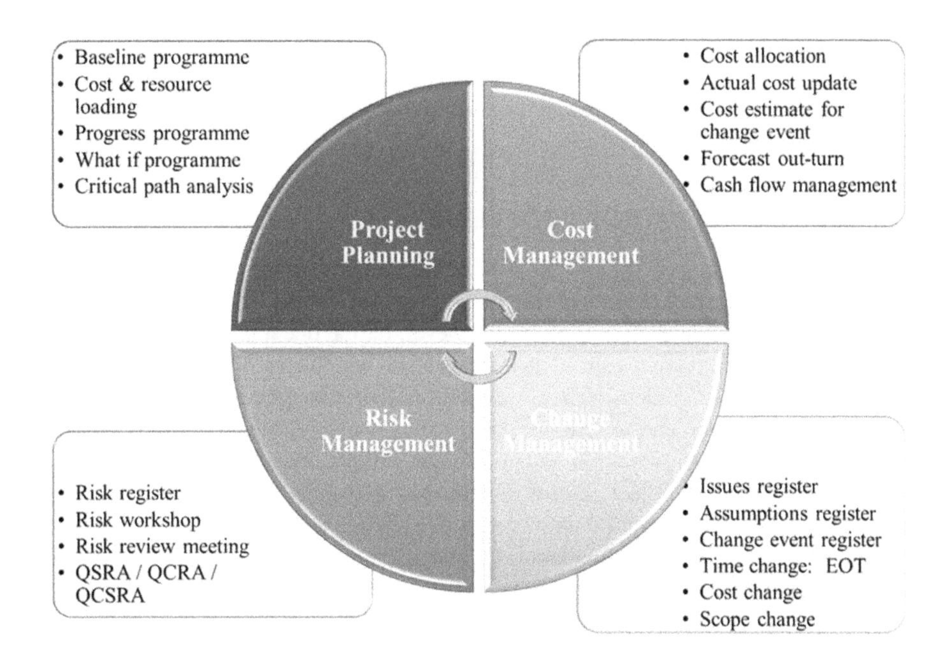

FIGURE 8.1 Integrated project control: key components and responsibilities.

updates the risk register. The project control team then integrates all parties' input into further analysis and feed back to each party.

Figure 8.1 explains the key components of an integrated project control system and the activities undertaken within each function. Together with progress reporting, lean construction and earned value management will be discussed further in Section 8.3.

8.2.2 Programme and Portfolio Management

Mega international infrastructure projects are generally implemented by a special purpose vehicle (SPV) limited company. For example, Crossrail Limited undertook the Crossrail project affiliated to TfL (Transport for London), and Riyadh Metro Limited undertook the Riyadh Metro project.

There are many main contractors who take part in the project components. Each main contractor will be associated with a number of sections and work packages. Therefore, the overall infrastructure project is actually a portfolio, and each main contractor constitutes the programme within the portfolio and the individual work package or service stream is the actual project. Consequently, in order to manage and control such mega infrastructure projects, it is also important to integrate the tools and methodology for the programme management and portfolio management into the overall integrated project control system.

For example, the PMI's standard for portfolio management and standard for programme management, as well the organizational project management maturity model (OPM3), provide detailed guidance for managing programme and portfolios that constitute multiple projects.

When developing the integrated project control system for large infrastructure projects or in the project management office within a large organization, the project control leader should also integrate control and management into the programme and portfolio level.

8.3 PROJECT CONTROL PROCEDURE

Many organizations view the project control narrowly as project planning plus project reporting. The project control function is separate from cost management, risk management, and change management. However, in practice it is more effective and efficient to treat the project control as a broad term, which covers time and cost control of the project's "golden triangle," as well as risk management, change control, contract management, and progress reporting. It is worth setting up the project organization structure with two main divisions under project management: the construction management and project control as described in Figure 8.2.

The project control function then constitutes project planning, cost management, change management, risk management, contract management, earned value management, and project reporting.

8.3.1 Earned Value Management

Earned value management plays an important role in project control. Almost every large construction project uses earned value management for project control. PMI (2011) describes earned value management as:

> *a management methodology for integrating scope, schedule, and resources; for objectively measuring project performance and progress; and for forecasting project outcome.*

The fundamental principle for earned value management is to compare the ongoing performance with the original baseline. It is well-recognized as an efficient tool for facilitating management and control project and indicating problem areas.

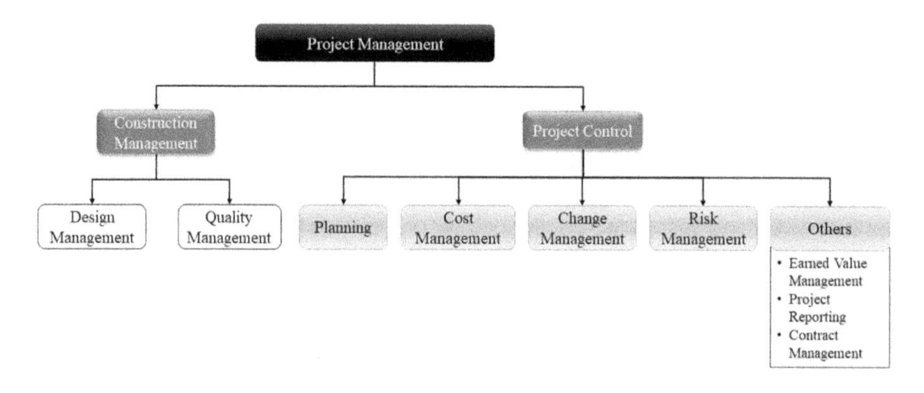

FIGURE 8.2 Recommended project structure for integrated project control.

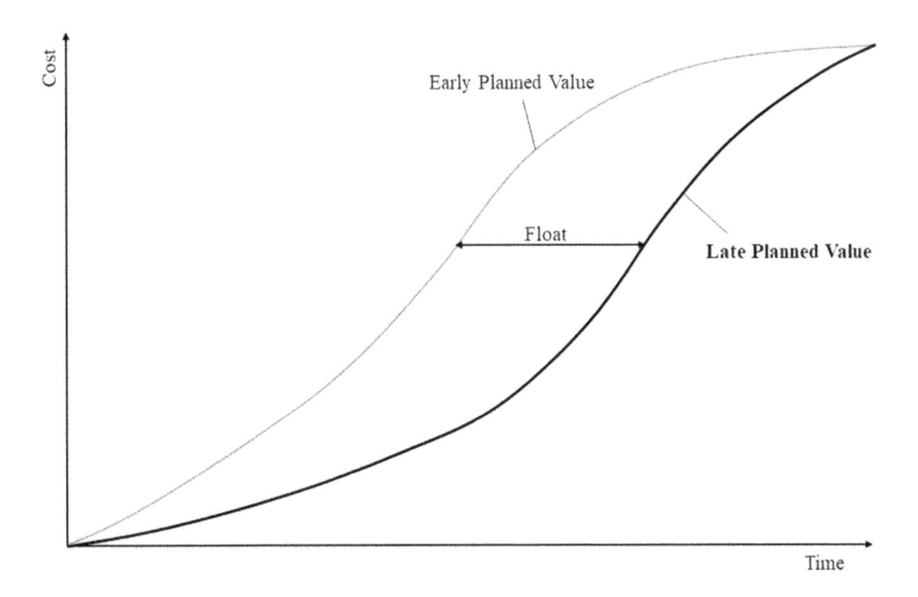

FIGURE 8.3 The banana chart: early planned value and late planned value.

In order to apply earned value management effectively and efficiently, it is better to have a resource and cost loading programme within the total project budget. The monthly budgeted cost can then be plotted out within the planning software, e.g., Primavera P6 or by exporting data in the Excel spreadsheet to prepare a separate analysis. Because each activity has an early Start/Finish date and late Start/Finish date, there are two planned values available for a single time point, namely the early planned value and late planned value as illustrated in Figure 8.3.

The gap between the two curves reflects the float in the project. This chart is often known as a banana chart. In practice, the early planned value is usually used as the planned value for earned value management. Therefore, the contractor is required to perform against the optimistic scenario of the project. Consequently, it is rare to see an SPI over 100%. If the SPI is around 90% or the SV is within the range of the total float, then there is no need for concern as regards project delay. In fact, it would be more appropriate to use the average planned value when undertaking earned value management, as described in Figure 8.4.

However, in order to establish a more accurate average planned value, the quality requirement of the programme is high. The logic link should be accurately established and have no open-ended activities other than the start milestone and finish milestone; although, using the average planned value gives the project manager and stakeholders more confidence in the performance indexes. In this approach, if the SPI is less than 100% or the SV is less than 0, potential delay is expected. Therefore, further analysis and mitigation need to be considered.

The programme is essential for earned value management, not only because it establishes the planned value from the baseline programme, but it is also the basis

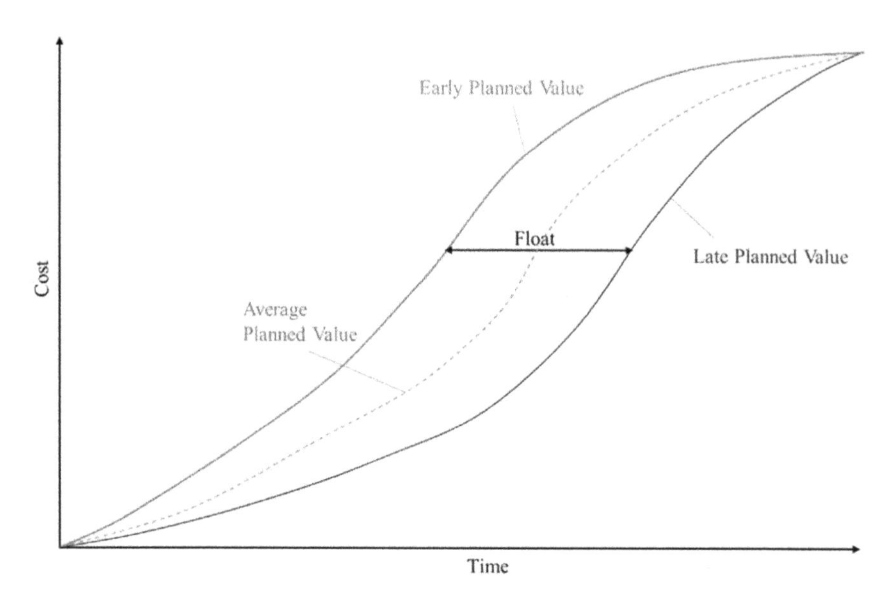

FIGURE 8.4 Average planned value.

for calculating the earned value, which is the cumulative value of the works completed to date, as shown in Equation 8.1.

$$\text{Earned Value} = \sum_{i=1}^{n} \text{Cost Activity Budgeted Cost}_i \times \text{Activity Physical \% Complete}_i.$$

Equation 8.1 Earned Value

Earned value analysis is undertaken by comparing the earned value with various components. Before discussing the earned value analysis tools in further detail, it is worth setting out the below abbreviations for the formulas:

- SPI – Schedule Performance Index
- CPI – Cost Performance Index
- SV – Schedule Variance
- CV – Cost Variance
- BAC – Budget at Completion
- AC – Actual Cost
- PV – Planned Value
- ETC – Estimate to Complete
- EAC – Estimate at Complete
- TCPI – To-Complete Performance Index

The most important component of earned value analysis is to understand the performance of cost and time at a certain cut-off point.

The performance of time is reflected in the schedule performance index (SPI) as shown in Equation 8.2 and schedule variance (SV) as shown in Equation 8.3.

$$SPI = \frac{EV}{PV}$$

Equation 8.2 Schedule Performance Index

$$SV = EV - PV$$

Equation 8.3 Schedule Variance

If the SPI is less than 100% or the SV is less than 0, then it means the project progress is behind the baseline programme. In contrast, if the SPI is greater than 100% or the SV is more than 0, then it means the project progress is ahead of the plan.

The performance of cost is illustrated in the cost performance index (CPI) as shown in Equation 8.4 and cost variance as shown in Equation 8.5.

$$CPI = \frac{EV}{AC}$$

Equation 8.4 Cost Performance Index

$$CV = EV - AC$$

Equation 8.5 Cost Variance

Likewise, if the CPI is less than 100% or the CV is less than 0, then it means the project progress is over the planned budget, and vice versa.

Equation 8.2 to Equation 8.4 illustrate that the SPI and/or the SV compare the earned value (EV) with the planned value (PV), whereas the CPI and/or CV compare the earned value (EV) with the actual cost (AC). Therefore, the earned value is key for analyzing both time and cost performance.

Both SPI/SV and CPI/CV can be calculated for the individual periods or the cumulated status to date. However, in order to undertake correct earned value analysis, the cut-off time must be the same for the planned value, earned value, and actual cost. The planned value and earned value are relatively easy to control with comprehensive planning software, e.g., Primavera P6. However, in practice, the actual cost often relates to the organization's financial cut-off date. It would be better to establish the same period end date for actual cost calculation or reporting as the data date for the programme update. The month end or the last day of four-weekly intervals is generally used as the cut-off date for project reporting. In the event that some of the sub-contractors' or suppliers' actual costs are not available by the time of reporting, the quantity surveyor should provide a forecast cost and update the correct actual value in the following period.

It is also worth noting that for projects undertaken with Options C or D of the NEC contract, the value of application for payment is often used as the actual cost in the earned value analysis. The amount of the *Contractor*'s application for payment constitutes the actual Defined Cost to the cut-off date and the forecast Defined Cost to the next assessment date. Therefore, it is important for the project controller to use the same time period for the actual cost, the planned value, and the earned value when undertaking earned value analysis under NEC options C and D and make necessary adjustment if required.

Apart from the time performance index and cost performance index, whether the project can be completed within the budgeted cost is another key concern for both the Client and the Contractor. Therefore, the cash flow forecast is often required as part of the project report.

The project outturn is usually calculated based on the actual cost to date plus the estimate to complete as described in Equation 8.6.

$$\mathrm{EAC} = \mathrm{AC} + \mathrm{ETC}$$

Equation 8.6 Estimate at Completion

With the actual cost to date fixed, the forecast cost of the remaining project becomes crucial to determine the end project outturn. Different contract types will use different methods to calculate the forecast cost. For the fixed price or lump sum contract, the estimate to complete can be simply calculated by the budgeted at completion cost (BAC) minus actual cost (AC) to date, as shown in Equation 8.7.

$$\mathrm{ETC} = \mathrm{BAC} - \mathrm{AC}$$

Equation 8.7 Estimate to Complete Under Lump Sum Contract

For the reimbursement contract, the estimate to complete is calculated in accordance with the cumulative cost performance index to date, as shown in Equation 8.8.

$$\mathrm{ETC} = \frac{\mathrm{BAC} - \mathrm{AC}}{\mathrm{CPI}}$$

Equation 8.8 Estimate to Complete Based on the CPI to Date

Because the delay of a project can also increase the total project cost, for some projects with high duration sensitivity under reimbursement contracts may use the combined cost and time performance index to date to determine the remaining project cost as shown in Equation 8.9.

$$\mathrm{ETC} = \frac{\mathrm{BAC} - \mathrm{AC}}{\mathrm{CPI}_{\mathrm{cum}} \times \mathrm{SPI}_{\mathrm{cum}}}$$

Equation 8.9 Estimate To-Complete Based on the CPI and SPI to Date

If the combination of the cumulated CPI and SPI still cannot reflect the appropriate projection for the forecast cost to complete, the proportion of each index can be considered. For example, use 70% of the CPI and 30% of the SPI. Alternatively, the project control manager can discuss with the planning manager and the cost manager to determine a fixed factor for each period to calculate the forecast cost.

In addition, the to-complete performance index (TCPI) is also used to predict the cost performance index for the remaining project by dividing the remaining works by the remaining project budget, as shown in Equation 8.10.

$$\mathrm{TCPI} = \frac{\mathrm{BAC} - \mathrm{EV}}{\mathrm{BAC} - \mathrm{AC}}$$

Equation 8.10 To-Complete Performance Index

With earned value management, the potential problem associated with the programme delay and cost overrun can be immediately identified in the earned value management S-Curve as demonstrated in Figure 8.5.

The schedule variance is indicated in the horizontal gap between the planned value and the earned value. Whereas the cost variance is indicated in the vertical gap between the earned value and the actual cost. Further, the horizontal gap between the end point of the planned curve and the estimate at completion curve illustrates the duration of delay to be expected. Meanwhile, the vertical gap between the end point of the planned curve and the estimate at completion curve minus the project contingency illustrates the potential cost overrun of the project.

8.3.2 Cost Control

Most of the Contractors, Consultants, Sub-contractors, and Suppliers spend much time and effort on cost management and control, because ultimately their objective when undertaking the project is to make a profit and maintain the operation of the business. Consequently, cost management and control often are undertaken separately and there is a limited connection to project control. In fact, in order to manage and

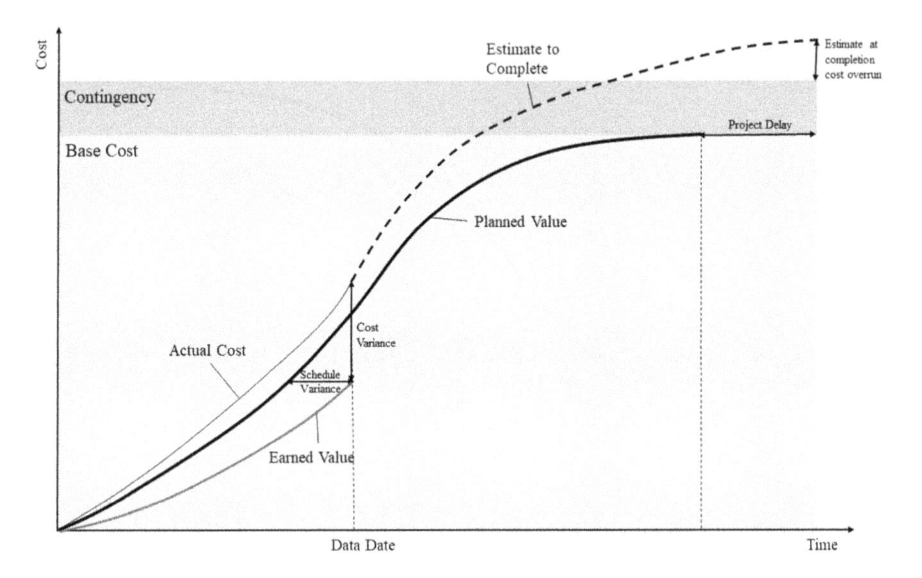

FIGURE 8.5 Earned value management S-curve.

control the cost effectively, appropriate programme need to be put in place to plan and manage the resource, forecast the cashflow, and maintain financial confidence.

In order to carry out efficient project control and earned value analysis, at the beginning of the project, it is particularly important to establish the appropriate linkage between the cost account set up in the contractor's financial system with the WBS programme or activity code. The commercial manager, planning manager, and project control manager should thoroughly consider setting up an appropriate cost account code before commencing the site works.

The contract cost establishes the starting point of the system. However, when undertaking the cost estimation during the tender stage, a different approach may be used. Ideally the bottom-up estimation is used as it can give a more accurate estimation of the total project cost. If the top-down estimation is undertaken, it is important to evaluate the cost allocation at the beginning of the project and provide resource and cost loading information for each activity. Once the total contract budgeted cost is allocated to the programme, the S-curve for the planned value is determined as the green line shown in Figure 8.6.

As the project progresses, change events arise and additional budget add to the project. As each change event is subject to agreement of the Client or its representative, there is a time lag for approving the change events. Once the Client approves the change request, the approved additional cost should be added to the existing project budget and the S-curve updated accordingly, as the purple line shown in Figure 8.6.

For some forms of contract, e.g., NEC and FIDIC, which requires the Contractor to carry on the work while waiting for the change approval, even if the change request is rejected by the Project Manager or the Engineer later on, the Contractor is entitled

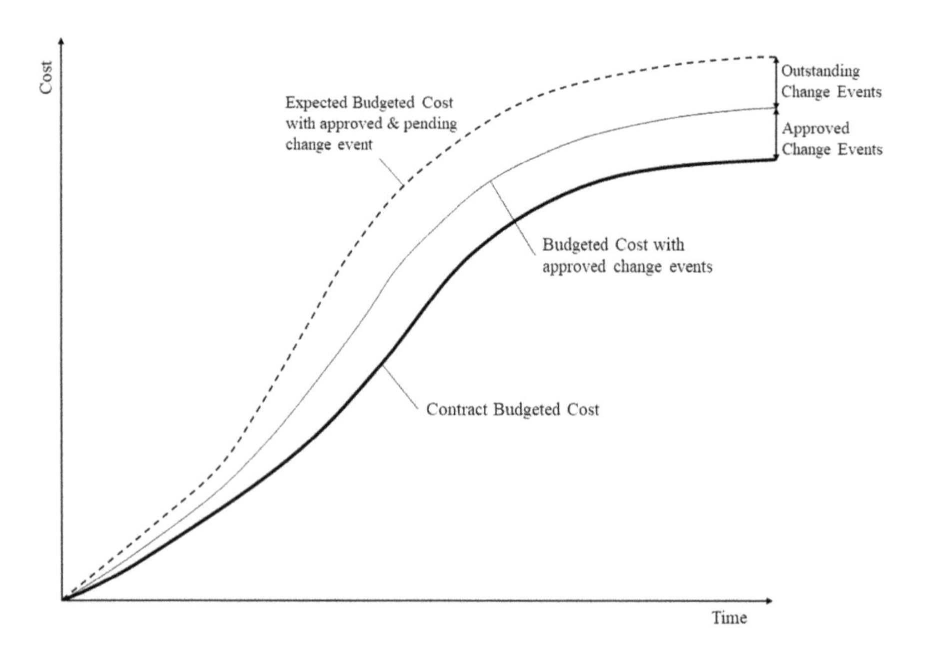

FIGURE 8.6 Different type of budgeted cost in earned value management.

to be paid for the actual cost it occurred for the proposed change event. In this case, it would also be worth tracking the performance against the expected budgeted cost with outstanding change events, as the dash line shows in Figure 8.6.

8.3.3 Schedule Control

Project planning is a fundamental component of earned value management as well as project control, no matter how narrowly or broadly defined. In order to establish the appropriate programme for effective earned value management, it is important to establish the standard planning components in the project control procedure at the beginning of the project.

Because the earned value is calculated based on the work completed to date, which is usually calculated based on the percentage completed multiplied by the total budgeted cost for each activity. It is important to update the correct percentage completed value in the programme. In Primavera P6, there are different types of percent complete, including the duration percent complete, physical percent complete, and unit percent complete. In practice, the task dependent activity is usually assigned with the physical percent complete, the unit percent complete is usually assigned to the resource dependent activity, and the level of effort (LoE) activity is generally assigned to duration percent complete. The earned value is generally calculated based on the physical percent complete. Therefore, when updating the programme, it is important to update the physical percent complete correctly for the types of activity defined as level of effort or resource dependent.

In addition, Primavera P6 provides three options to calculate the earned value as follows:

- At completion value with current dates
- Budgeted value with current dates
- Budgeted value with planned dates

In practice, the "Budgeted value with planned dates" option is generally selected for the automatic earned value analysis.

The project control manager should set out the standard earned value calculation options in the project control procedure to ensure a consistent approach is used for the earned value analysis, particularly for large infrastructure projects which constitute multiple projects and multiple programme.

When assigning the baseline programme to the progress programme, Primavera P6 can automatically calculate the earned value and show the SPI for each activity in a separate column. As discussed in Section 8.3.1, by default, the early planned value has been used in the earned value analysis. If the project chooses to use the reasonable average planned value as described in Figure 8.4, the project control manager can set up a standard global change with the formulas as described in Equation 8.11, to calculate the adjusted SPI based on the average planned value.

$$SPI_{adj.} = \frac{EV}{\frac{PV_{early} + PV_{late}}{2}}$$

Equation 8.11 Schedule Performance Index Based on Average Planned Value

Likewise, the application of such global change formulas should be clearly stated in the project control procedure in order to keep consistency of the overall analysis across all work packages and projects.

Similar to cost control, for each approved change event, the approved programme needs to be integrated into the main programme. For projects undertaken in the traditional method, where the baseline programme for the original contract scope would be fixed throughout the project, the activities associated with the approved change event need to be updated in the baseline programme as well. Meanwhile, resource and cost should allocate the relevant activities in accordance with the approved change cost. It is important to ensure the same activity ID is used in the baseline programme and all other versions of progress programme, otherwise the P6 integrated earned value analysis will not work accurately. For projects undertaken under the NEC or FIDIC contracts, it also worth setting up a separate WBS at the bottom of the project to show the programme associated with the pending changes. Once the pending change event is approved by the Project Manager or the Engineer, it can then be incorporated into the main programme.

In addition to the programme being managed in planning software, e.g., Primavera P6, the site manager may also keep a separate daily programme to control the daily activities on site. Such a detailed short-term programme is usually called a work plan. These work plans are too detailed to be included in the overall project programme, but it is important to have such a short-term detailed work plan in an Excel or Microsoft Project format to facilitate the site manager to control the daily activities on site. Again, the project control manager should set out the standard template and requirement for the weekly work plan in the project control procedure.

8.3.4 Change Control

Apart from updating the cost associated with changes to cost management and the programme, the change control team also need to keep clear records of the status and features of each change event. For example, for projects under the NEC contract, the compensation event should be maintained and reported each period for the rejected compensation events, the compensation events to be quoted, the approved compensation events, the compensation events subject to the *Project Manager*'s assessment, and the compensation events to be approved should be reported for both the current period and cumulated to date for the number of compensation events and the value associated with these compensation events, as shown in Table 8.1.

Due to the complexity of large construction projects, change control can be a very challenging task. Each individual change event needs to comply with the process and time limit required by the contract. In addition, each change event needs to provide a quotation and/or programme and be submitted for approval. Once the proposed change event is approved, then all the relevant cost and programme need to be updated accordingly. It would be better to use an existing software, e.g., CIMAR for the NEC contract, or develop a change control system at the front-end of the project to facilitate the change control team to follow the process, provide reminders for the time due, and prepare reports for various analyses and control.

TABLE 8.1
Compensation Event Control and Report

	No.			Value (£)		
Compensation events	**Previous**	**Period**	**Cum**	**Previous**	**Period**	**Cum**
CEs rejected						
CEs to be quoted						
CEs to be approved						
Project Manager's Assessment						
CEs Agreed						
Total CEs						

8.3.5 Risk Control

Risk control can be undertaken in two routes. For risks that have a low to medium impact, the control is usually undertaken within the risk register; the frequent risk review meeting is undertaken and the mitigation actions are discussed and assigned to the competent person.

It is also important to create a risk friendly culture across the project, so that the project participants can raise the potential risks to the risk manager as the project progresses.

For the risks that may have a significant impact, the comprehensive quantitative schedule risk analysis and quantitative cost risk analysis is undertaken to facilitate the decision making for the senior management between both contract parties.

In theory, as long as the earned value curve is within the banana curve as shown in Figure 8.3, the project should still be able to achieve the targeted completion date. In the circumstance of the early planned value having been used for the earned value analysis, if the SPI is less than 100%, but the project manager would like to achieve confidence in the project completion date, a quick QSRA with duration uncertainty can be carried out; the confidence level of achieving the completion date can then provide further clarification. If certain activities with high sensitivity in the tornado chart impact the project completion date or certain key dates, then more focus should be given to them.

In addition, it is also important to control drawdown of risk contingency. The relevant process should also be clearly set out in the project control procedure or the risk management plan at the beginning of the project. The top risks and opportunities together with the high-level risk values should also be reported in the progress report. Figure 8.7 provides an example of reporting the risk and opportunity management status in the periodical progress report.

8.3.6 Performance Reporting

Performance reporting in large-scale construction projects can involve substantial works. It is often the case that the detailed monthly report can end up between 100

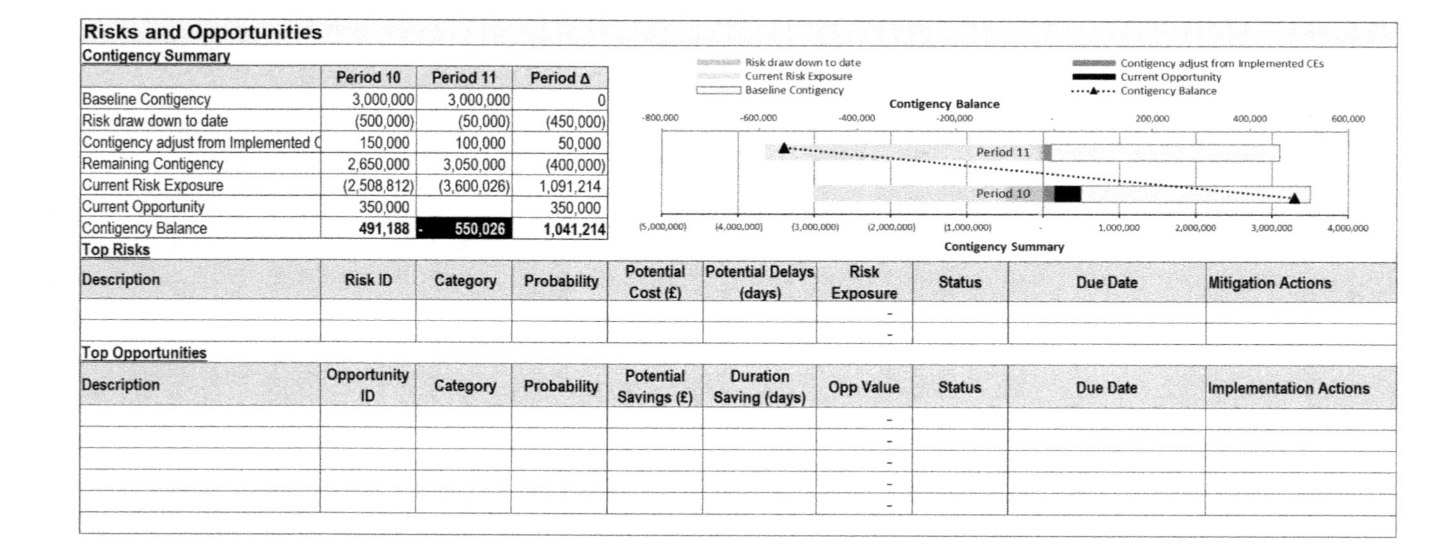

Risks and Opportunities

Contigency Summary

	Period 10	Period 11	Period Δ
Baseline Contigency	3,000,000	3,000,000	0
Risk draw down to date	(500,000)	(50,000)	(450,000)
Contigency adjust from Implemented C	150,000	100,000	50,000
Remaining Contigency	2,650,000	3,050,000	(400,000)
Current Risk Exposure	(2,508,812)	(3,600,026)	1,091,214
Current Opportunity	350,000		350,000
Contigency Balance	**491,188**	- 550,026	**1,041,214**

Top Risks

Description	Risk ID	Category	Probability	Potential Cost (£)	Potential Delays (days)	Risk Exposure	Status	Due Date	Mitigation Actions
						-			
						-			

Top Opportunities

Description	Opportunity ID	Category	Probability	Potential Savings (£)	Duration Saving (days)	Opp Value	Status	Due Date	Implementation Actions
						-			
						-			
						-			
						-			
						-			

FIGURE 8.7 Example of risk and opportunity report.

and 300 pages. In addition, various reports need to be prepared for different purposes, e.g., the progress report for the client, the internal progress report for the contractor's Joint Venture board, the KPI report for Option X20 under the NEC contract, the overall couple of pages dashboard for the contract board, etc.

In order to prepare and submit all the different reports required on time, it is important to establish the linkage between the various reports and set up the standardized template.

For example, the template sets out the required data input for the earned value analysis. In order to establish automatic data transfer to the pre-established template, the data export template from Primavera P6 and the contractor's financial system must be pre-established and kept consistent.

A typical progress report starts with the executive summary to provide the overall progress of the performance in the period and highlight the strategic actions to be undertaken in order to achieve the project objectives or to mitigate key issues or risks. The detailed sections included in the progress report generally include health and safety, performance analysis with the earned value management, programme status, cost management, change management associated with key issues, and risk and opportunity management, design, and quality management. Some projects also include sections on environment management, stakeholder management, the site photos to demonstrate the physical progress and achievement on site, etc. It is important to agree the content, format, and level of detail of the progress report at the beginning of the project. Once agreed, the project control manager should consider establishing a relevant process for each component of reporting in the project control procedure.

It is also important to set up the standard reporting calendar for each contributor, including the cut-off date for actual cost, the data date of the planning, the deadline for the planner and quantity survey or to complete their programme and cost update, the deadline for each section contributor to produce their input, the senior management review date, as well as the final report submission date.

The earned value analysis is one of the most important components in the progress report. As discussed in Section 8.3.1, it is based on the various values presented in the S-curve for a certain reporting point. Figure 8.8 provides an example of the S-curve chart to demonstrate the project performance to date in each period as well as the cumulative position.

In accordance with the resource profile and the earned value achieved in the past period, a trend of the Contractor's performance can be worked out. If the forecast value in the remaining periods is significantly higher than the average earned value in the past periods, the project manager can use the earned value trend line to project the realistic project completion date and challenge the contractor's programme.

The earned value analysis should be undertaken in the different sections and further breakdown to cost account level within each section. In this way, the project manager obtains an overview of the overall project performance; if any EVM indicators show signs of warning, the project manager can drill down to the specific section and further to the relevant cost account.

Furthermore, under the NEC contract, the control and report for Defined Cost, Disallowed Cost, and Final forecast outturn are often required. Therefore, the project

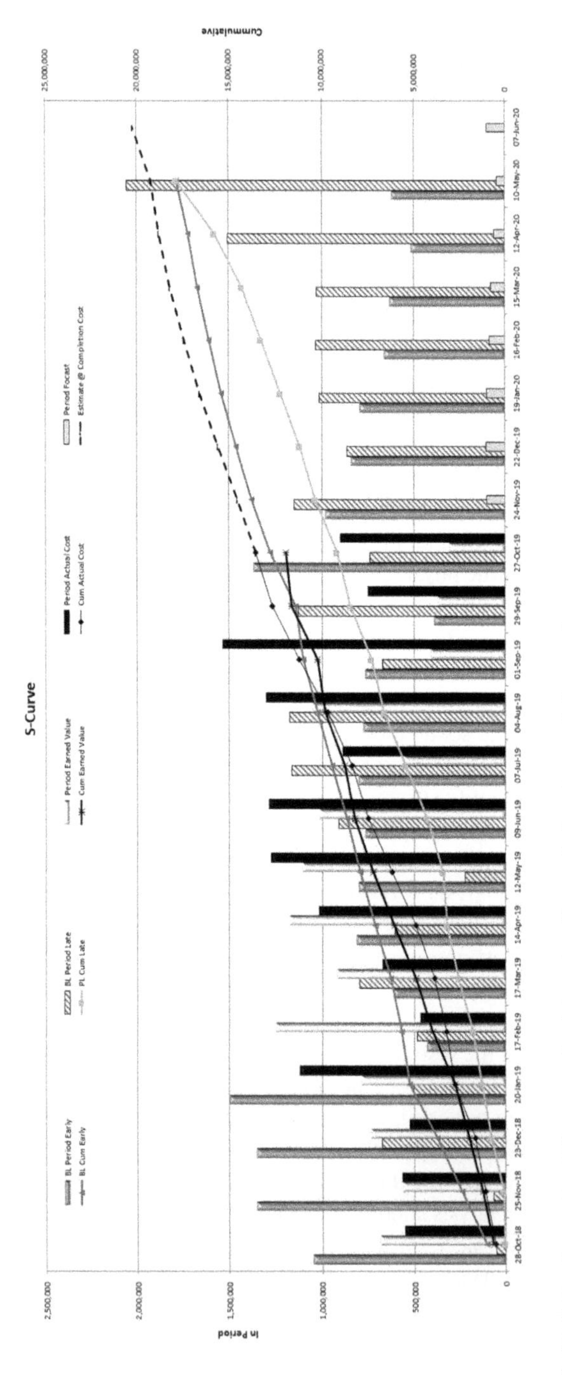

FIGURE 8.8 Example of S-curve chart showing the period and cumulative performance.

control manager should discuss with the cost manager to ensure an appropriate procedure is set up to capture these components in the period progress report.

In addition, if the X20 "Key Performance Index" is used in the NEC contract, the project control manager should also set up a separate KPI report for the elements required under Option X20 to present the achievements in each period and automatically establish a relevant reward in accordance with the contract.

8.4 CONTINUOUS PERFORMANCE IMPROVEMENT

Due to the complexity of large-scale construction project, in order to successfully control and deliver the project, it is important to adapt the control process and tools to the dynamic changing situations. Various incentives set out in the standard forms of contract, e.g., KPI (Option X20 under NEC), value engineering (Clause 63.12 under NEC4 and Sub-Clause 13.2 under FIDIC 2017), whole life cost (Option X21 under NEC4) are all encourage the Contractor to improve its performance continuously.

8.4.1 Key Performance Indicators (KPI)

Option X20 "Key Performance Index" under the NEC contract provides incentives to encourage the continuous improvement of the *Contractor*'s performance. Such incentives can also be set up between the *Contractor* and its Sub-contractor. The KPI can include the following elements:

- key health and safety indicators, e.g., zero RIDDAR or dangerous occurrence in the period
- the schedule performance index
- the cost performance index
- the quality indicators, e.g., outstanding defects and percentage of defects corrected
- the accuracy of the forecast

An incentive payment is defined in the contract. If the *Contractor* achieves the target KPIs then the incentive will be paid in the next period. Such an incentive payment does not count towards the Total of the Prices. Therefore, it is a net gain to the *Contractor* and will not be subject to the Pain/Gain share if the NEC Options C or Option D are used.

The Parties can agree additional performance measures in accordance with the requirement of the project during the implementation. This allows a continuous incentive to encourage the *Contractor* to improve its performance.

8.4.2 Lean Construction

The construction industry has very low profit margin. The main contractor usually has less than 5% profit margin and the consultant usually has around 10% profit margin. Yet, the waste in construction projects is generally far more than the contractor's profit. In order to achieve project success, most tier one contractors now pay more attention to lean construction. The typical lean technique uses a DMAIC (Define, Measure, Analyze, Improve, and Control) approach to improve the management process. Once the potential issue is defined, the existing performance measures will be undertaken, and the quick wins will be identified. The analysis of types of wastes against the seven typical wastes under the Six-Sigma or lean techniques will be carried out. Based on the type of wastes, the root cause analysis will then be undertaken and the 5 Whys tool is often used to find out the root cause. Having understood the root cause, the corrective action to improve the process will be developed. If necessary, the new process map will be developed and finalized with the FMEA (Failure Model Effect Analysis) and cost benefit analysis. The most well-used lean tool on construction sites includes the 5S (Sort, Set, Shine, Standardize, Sustain) to improve the site order. Value stream mapping is often used to determine the waste associated with the material and resource. Finally, the control plan will be developed and reviewed to ensure improvement action is taken and the process is continuously improved and waste generation is limited.

The tools more relevant to project control is collaborative planning and the Last Planner®. The interactive planning workshop discussed in Section 8.3.1 is a type of collaborative planning. In addition, the collaborative planning is undertaken throughout the project implementation. In particular, the site manager often uses it to plan the detailed activities with the interface parties before work starts each week in order to clearly identify the constraints and optimize the construction sequence with the interface parties. In recent years, the Last Planner® System, developed by the Lean Construction Institute, has achieved success in the onsite planning of daily works. As the implementation plan is developed by the person who actually undertakes the physical work on site, it is more realistic and more reliable.

Lean techniques are certainly a powerful tool to facilitate the continuous improvement of the project control processes and eventually lead to the project success.

8.5 RECOMMENDATIONS

An integrated project control system is important to project success. It is recommended for both consultancy and construction organizations establish an integrated project control system at the corporate level, which can be tailored and adapted to various projects. It is recommended for organizations that already have the Project Management Office (PMO) in place, to also shift their attention to establishing an appropriate integrated project control system.

For large infrastructure projects, which are often undertaken with a specific project limited company, it is highly recommended that the senior management team recruit competent personnel to establish an appropriate integrated project control system and establish an appropriate project control procedure before the project commencement. Then the client should require all the contractors and the sub-contractors to comply with the project control procedure. It would be good practice to include incentive and sanction clauses within the contract. For example, Option X20 "Key Performance Index" under the NEC contract, the sanction of failure to submit the first programme on time under Clause 50.5 of the NEC4 contract, and the time bar for the variation and/or compensation event under the FIDIC and NEC contracts respectively.

With the development of information technology, there are many tools and softwares that can facilitate organizations to establish an integrated project control system. For example, Oracle has now shifted their focus from developing individual products for project planning (Primavera P6), risk management (Primavera Risk Analysis), and document control (Aconex), to developing an integrated project management control system "PRIME," which is a web-based system. This new system integrates its existing successful products including the Primavera P6 planning function, the risk analysis and management function of the Primavera Risk Analysis, the document control under Aconex, together with Primavera Lean and Primavera Portfolio to provide organizations that undertake large construction projects with a fully integrated platform to manage and control their project in a dynamic environment.

Having established the integrated project control system, having the personnel to correctly and effectively use the system to undertake appropriate project control then becomes important. The organization should appoint the competent personnel to lead the project control function, including project planning, change management, contract management, cost management, and risk management. In addition, a personal development plan should be in place to identify potential gaps within the project control team and provide relevant and timely training.

REFERENCES

Book/Article

Logikal. 2019. *2019 Project Controls Survey Report*. Logikal Project Intelligence.

PMI. 2011. *Practice Standard for Earned Value Management*. 2nd Edition. Newtown Square, PA: Project Management Institute.

Contract

FIDIC. 1999. *Conditions of Contract for Construction*. 1st Edition. (1999 Red Book). Geneva, Switzerland: The Fédération Internationale des Ingénieurs-Conseils.

FIDIC. 2017. *Conditions of Contract for Construction*. 2nd Edition. (2017 Red Book). Geneva, Switzerland: The Fédération Internationale des Ingénieurs-Conseils.

NEC. 2017. *NEC4 Engineering and Construction Contract*. London, UK: Thomas Telford Ltd.

Index

For Product Safety Concerns and Information please contact our EU representative GPSR@taylorandfrancis.com
Taylor & Francis Verlag GmbH, Kaufingerstraße 24, 80331 München, Germany

www.ingramcontent.com/pod-product-compliance
Ingram Content Group UK Ltd.
Pitfield, Milton Keynes, MK11 3LW, UK
UKHW021119180425
457613UK00005B/150

* 9 7 8 1 1 3 8 3 8 9 3 3 5 *